KEYS TO SOIL TAXONOMY

by

Soil Survey Staff

Agency for International Development

United States Department of Agriculture

Soil Management Support Services

SMSS Technical Monograph #6

Third Printing, 1987
Cornell University

KEYS TO SOIL TAXONOMY

Prepared by Agronomy Department
Cornell University
Ithaca, NY

For

THE SOIL MANAGEMENT SUPPORT SERVICES

Soil Management Support Services is a program of international technical assistance in soil survey, classification, interpretation, and use and management of soils in the intertropical countries of the Agency for International Development implemented by the Soil Conservation Service of the United States Department of Agriculture under PASA No. AG/DSB-1129-5-79.

This publication was produced under a cooperative agreement between the Soil Management Support Services (SMSS) and the College of Agriculture and Life Sciences at Cornell from funding by the Agency for International Development (A.I.D.).

The contents of this publication were provided by the Soil Conservation Service (SCS). The Agronomy Department prepared the camera-ready manuscript from text-files provided by SCS, rearranged some sections, produced the index, and converted measurement units into International System of Units (SI) terms.

To obtain additional copies in developing countries:

Ask your AID country mission or write to:
SMSS Program Leader
Soil Conservation Service
U.S. Department of Agriculture
P.O. Box 2890
Washington, DC 20013

To obtain additional copies in the USA and developed countries write to:
International Soils
Department of Agronomy
Bradfield Hall
Cornell University
Ithaca, NY 14853

US and developed countries' requests should include a check for $12.00 in US dollars per copy made out to the Agronomy Dept., Cornell University.

Citation:
Soil Survey Staff. 1987. Keys to Soil Taxonomy (third printing). SMSS technical monograph no. 6. Ithaca, New York.

Third Printing
1987

ISBN 0-932865-10-0

iii

Table of Contents

Foreword

This publication, *Keys to Soil Taxonomy*, serves two purposes. It provides the taxonomic keys required for the classification of a soil in a form that can be used easily in the field, and it serves as a means of providing an up-to-date version of *Soil Taxonomy* that includes all revisions to the keys that have been approved. It replaces the keys in Soil Taxonomy, but it does not replace Agriculture Handbook 436, which contains descriptive material, laboratory data, and chapters on other subjects related to Soil Taxonomy. This is the second update of the *Keys to Soil Taxonomy*, which was originally printed in 1983, and we plan to continue reprinting the keys about every two years. Agriculture Handbook 436 will be revised and reissued in the future, after most of the current ongoing International Soil Classification Committees (ICOMs) have completed their mandates, but probably not before about 1995.

This publication incorporates all amendments approved to date and published in National Soil Taxonomy Handbook (NSTH) Issues 1 through 11. It includes the recommendations of the International Committee on Low Activity Clays (NSTH Issue #8) and the International Committee on Oxisols (NSTH Issue #11). The keys reproduced here were extracted from a computerized copy of *Soil Taxonomy* which is maintained in complete, up-to-date form.

RICHARD W. ARNOLD
1987
Director, Soil Survey Division
Soil Conservation Service

Chapter 1

The Soil That We Classify and Horizons and Properties Diagnostic for the Higher Categories: Mineral Soils

The Soil That We Classify

Soil, as used in this text, is the collection of natural bodies on the earth's surface, in places modified or even made by man of earthy materials, containing living matter and supporting or capable of supporting plants out-of-doors. Its upper limit is air or shallow water. At its margins it grades to deep water or to barren areas of rock or ice. Its lower limit to the not-soil beneath is perhaps the most difficult to define. Soil includes the horizons near the surface that differ from the underlying rock material as a result of interactions, through time, of climate, living organisms, parent materials, and relief. In the few places where it contains thin cemented horizons that are impermeable to roots, soil is as deep as the deepest horizon. More commonly soil grades at its lower margin to hard rock or to earthy materials vitually devoid of roots, animals, or marks of other biologic activity. The lower limit of soil therefore is normally the lower limit of biologic activity, which generally coincides with the common rooting depth of native perennial plants. If either biological activity or current pedogenic processes extend to depths greater than 2 m, the lower limit of the soil we classify is arbitrarily set at 2 m. Yet in defining mapping units for detailed soil surveys, lower layers that influence the movement and content of water and air in the soil of the root zone must also be considered.

Buried Soils

A soil is considered to be a buried soil if there is a surface mantle of new material that is 50 cm or more thick or if there is a surface mantle between 30 and 50 cm thick and the thickness of the mantle is at least half that of the named diagnostic horizons that are preserved in the buried soil. A mantle that is <30 cm thick is not considered in the taxonomy but, if important to the use of the soil, is considered in establishing a phase. The soil that we classify in places where a mantle is present, therefore, has its upper boundary at the surface or <50 cm below the surface, depending on the thickness of its horizons.

A surface mantle of new material as defined here is largely unaltered. It is usually finely stratified and overlies a horizon sequence that can be clearly identified as the solum of a buried soil in at least part of the pedon, as defined in the following chapter. The recognition of a surface mantle should not be based solely on studies of associated soils.

Mineral Soil Material

Mineral soil material either

1. Is never saturated with water for more than a few days and has <20 percent organic carbon by weight; or

2. Is saturated with water for long periods or has been artificially drained, and has
   a. Less than 18 percent organic carbon by weight if 60 percent or more of the mineral fraction is clay;
   b. Less than 12 percent organic carbon by weight if the mineral fraction has no clay; or
   c. A proportional content of organic carbon between 12 and 18 percent if the clay content of the mineral fraction is between zero and 60 percent.

Soil material that has more organic carbon than the amounts just given is considered to be organic material.

Definition Of Mineral Soils

Mineral soils, in this taxonomy, are soils that meet one of the following requirements:

1. Mineral soil material <2 mm in diameter (the fine-earth fraction) makes up more than half the thickness of the upper 80 cm (31 in.);

2. The depth to bedrock is <40 cm and the layer or layers of mineral soil directly above the rock either are 10 cm or more thick or have half or more of the thickness of the overlying organic soil material; or

3. The depth to bedrock is $\geq$40 cm, the mineral soil material immediately above the bedrock is 10 cm or more thick, and either
   a. Organic soil material is <40 cm thick and is decomposed (consisting of hemic or sapric materials as defined later) or has a bulk density of 0.1 g cm^{-3} or more; or
   b. Organic soil material is <60 cm thick and either is undecomposed sphagnum or moss fibers or has a bulk density that is <0.1.

Diagnostic Surface Horizons; The Epipedon

Six diagnostic horizons that form at the surface are defined. Any horizon, however, may be at the surface of a truncated soil. A horizon that forms at the surface is called an epipedon (Gr. *epi*, over or upon, and *pedon*, soil). The epipedon not only has formed at the surface but it also has been either appreciably darkened by organic matter or eluviated or, as a minimum, rock structure has been destroyed. Such a horizon may become covered by thin deposits of fresh alluvium or by thin eolian deposits without losing its identity as an epipedon. The depth to which an epipedon must be buried to be considered part of a buried soil is defined below. Generally a buried horizon lies below a depth of 50 cm or more, usually more.

There can be only one epipedon formed in the mineral surface horizon(s) of a soil. This epipedon may be overlain by organic materials that may meet the definition of a histic epipedon (defined later). Otherwise one soil may contain only one epipedon.

A recent alluvial or eolian deposit that retains fine stratifications or an Ap horizon that is directly underlain by material that retains fine stratifications are not are not included in the concept of the epipedon because time has not been sufficient for soil-forming processes to erase these transient marks of deposition and for diagnostic and accessory properties to develop.

The epipedon is not a synonym for the A horizon because it may include part or all of the illuvial B horizon if the darkening by organic matter extends from the surface into or through the B horizon. To avoid changes in classification of a soil as the result of plowing, the properties of the epipedon, except for structure, should be determined after the surface soil to a depth of 18 cm has been mixed or, if the depth to bedrock is <18 cm, after the whole soil down to rock has been mixed.

Anthropic epipedon

In summary, the anthropic epipedon conforms to all the requirements of the mollic epipedon except (1) the limits on acid-soluble P_2O_5, with or without the base saturation, or (2) the length of the period during which it has available moisture. Additional data on anthropic epipedons from several parts of the world may permit improvements in this definition.

Histic epipedon (Gr. *histos*, tissue)

The histic epipedon normally is at the surface, although it may be buried at a shallow depth. It normally is a thin horizon of peat or muck if the soil has not been plowed. If the soil has been plowed, the histic epipedon has the very high content of organic matter that results from mixing peat with some mineral material. Since peaty deposits occur in wet places, the histic epipedon either is saturated with water for 30 consecutive days or more during the year or has been artificially drained.

The histic epipedon, therefore, can be defined as a layer (one horizon or more) at or near the surface that is saturated with water for 30 consecutive days or more at some time in most years, or is artificially drained, and that meets one of the following requirements:

1. The surface horizon consists of organic soil materials that either
   a. Is 75 percent or more, by volume, sphagnum fibers or has a bulk density, when moist, <0.1 and is <60 cm (24 in.) but >20 cm thick; or
   b. Is <40 cm but >20 cm thick and meets one of the following requirements with respect to organic-carbon content and thickness:
      (1) Has 18 percent or more organic carbon if the mineral fraction is 60 percent or more clay;

(2) Has 12 percent or more organic carbon if the mineral fraction has no clay;
(3) Has an intermediate proportional content of organic carbon if part but less than 60 percent of the mineral fraction is clay.

2. The plow layer is 25 cm or more thick and has 8 percent or more organic carbon if it has no clay, or 16 percent or more organic carbon if 60 percent or more of the mineral fraction is clay, or an intermediate proportional content of organic carbon if part but less than 60 percent of the mineral fraction is clay.

3. A layer of organic material that has enough organic carbon and is thick enough to satisfy one of the requirements under item 1 lies beneath a surface layer of mineral materials that is <40 cm (16 in.) thick. In such a soil, the histic epipedon has been buried but the mineral materials at the surface are too thin to be considered diagnostic in the classification.

4. A surface layer of organic material <25 cm thick that has enough organic carbon to satisfy the minimum requirements under item 2 after the soil has been mixed to a depth of 25 cm.

Mollic epipedon (L. *mollis*, soft)

The mollic epipedon is defined in terms of its morphology rather than its genesis. It consists of mineral soil material. It is a surface horizon or horizons unless (a) it underlies a recent deposit that is <50 cm thick and that has fine stratifications if not plowed or (b) it underlies a thin layer of organic material in a wet soil (see histic epipedon). If the layer of organic material is thick enough that the soil is organic, the mineral soil is considered to be buried.

The mollic epipedon has the following properties:

1. Soil structure is strong enough that the major part of the horizon is not both massive and hard, or very hard when dry. Very coarse prisms, >30 cm in diameter, are included in the meaning of massive if there is no secondary structure within the prisms.

2. Unless there is >40 percent finely divided lime, both broken and crushed samples have Munsell color value darker than 3.5 when moist and 5.5 when dry, and chroma less than 3.5 when moist[1]; the color value normally is at least 1 Munsell unit darker or the chroma is at least 2 units less (both moist and dry) than that of the 1C horizon if an 1C horizon is present. If only a 2C horizon or an R layer is present, the comparison should be made with the horizon that overlies the 2C. Some parent materials such as loess, cinders, alluvium, or carbonaceous shales also can have dark color

---

[1] The chroma is permitted to range up to but not to include 4.0 soils that have a hyperthermic or isohyperthermic temperature regime. The color when moist is that of a specimen that is moist enough that an additional drop of water produces no change in the color. The color when dry is that of a specimen dry enough that continued drying produces no further change.

and low chroma. Soils formed in such materials may accumulate appreciable amounts of organic matter but have no visible darkening in the epipedon. In this situation, the requirement that the mollic epipedon have lower value or chroma than the 1C horizon, or than the next underlying horizon if there is no IC, is waived if (a) the surface horizon (horizons) meets all other requirements for a mollic epipedon and, in addition, has at least 0.6 percent more organic carbon than the 1C or the 2C horizon or if (b) the epipedon extends to rock (either a lithic or paralithic contact as defined later).

The mollic epipedon is expected to have dark color and low chroma throughout the major part of its matrix. If the structure is fine granular or fine blocky, the color when broken may be only the color of the coatings. The color of the matrix in such situations can be determined only by crushing or briefly rubbing the sample. Prolonged rubbing should be avoided because it may cause darkening of a sample if soft iron-manganese concretions are present; crushing should be only enough to break and mix the coatings. The color value when dry should determined after the crushed sample has been smoothed to eliminate shadows.

If there is >40 percent finely divided lime, the limits of color value, dry, are waived; the color value, moist, then should be 5 or less. This waiver is necessary because finely divided lime acts as a white pigment.

3. Base saturation is 50 percent or more by the NH_4OAc method.

4. The organic-carbon content is 2.5 percent or more in the upper 18 cm if the color requirement is waived because of finely divided lime. Otherwise, the organic-carbon content is at least 0.6 percent (1 percent organic matter) throughout the thickness of soil specified in item 5.

The mollic epipedon consists of mineral rather than organic soil material. Its organic-carbon content, therefore, has an upper as well as a lower limit. The upper limit of organic carbon in a mollic epipedon is the same as that of mineral soil material; in part it is the lower limit for the histic epipedon, defined later in this chapter. Because an organic horizon can form above a mollic epipedon in a wet soil, the mollic epipedon is not necessarily the surface horizon but is the uppermost horizon composed of mineral soil material.

5.The thickness is one of the following after mixing the upper 18 cm of soil or the whole soil if the depth to rock, petrocalcic horizon, or duripan is <18 cm:
    a. Ten cm or more if the epipedon is underlain directly by a lithic contact; 10 cm or more in soils of shallow families in which the epipedon is underlain directly by a paralithic contact, a petrocalcic horizon, or a duripan, all defined later in this chapter;
    b. In other soils the epipedon must be >25 cm thick if its texture if finer than loamy fine sand and
        (1) The upper boundary of pedogenic lime that is present as filaments, soft coatings, or soft nodules is deeper than 75 cm;

(2) The base of any argillic, natric, spodic, cambic, or oxic horizon is deeper than 75 cm; and

(3) The upper boundary of any petrocalcic horizon, fragipan, or duripan is deeper than 75 cm;

c. In other soils that have a loamy or clayey epipedon, the thickness of the epipedon must be 18 cm or more and it must be more than one-third of the depth from the top of the epipedon to the shallowest of one of the features listed in (b) if that is <75 cm;

d. In other soils the epipedon must be >25 cm thick if

(1) The texture of the epipedon is as coarse as or coarser than loamy fine sand throughout its thickness or

(2) If there are no underlying diagnostic horizons and the organic-carbon content of the underlying materials decreases irregularly with increasing depth (as in recent alluvium that is not finely stratified); or

e. In other soils, the epipedon must be 18 cm (7 in.) or more thick if none of the conditions that are listed in *b*, *c*, and *d* exist.

6. The epipedon has <250 mg kg^{-1} of P_2O_5 soluble in 1 percent citric acid or it either has increasing amounts of P_2O_5 soluble in citric acid below the epipedon or the amounts of P_2O_5 soluble in citric acid decrease or increase irregularly with depth below the epipedon, or there are phosphate nodules within the epipedon. This restriction is made to eliminate plow layers of very old arable soils and kitchen middens that have acquired, under use, the properties of the mollic epipedon but to include the epipedon of a soil developed in highly phosphatic parent material.

7. If the soil is not irrigated, some part of the epipedon is moist 3 months or more of the year (cumulative) in more than 7 out of 10 years at times when the soil temperature at a depth of 50 cm is 5°C or higher.

8. The *n* value (defined later in this chapter) is <0.7. Although many soils that have a mollic epipedon are very poorly drained, a mollic epipedon does not have the very high water content of sediments that have been continuously under water since deposition.

Ochric epipedon (Gr. *ochros*, pale)

An ochric epipedon is one that is too high in value or chroma, is too dry, has too little organic matter, has an *n* value too high, or is too thin to be mollic, umbric, anthropic, plaggen or histic, or it is both hard and massive when dry. An epipedon is ochric if the Munsell color value after rubbing is 5.5 or higher when dry of 3.5 or higher when moist, if the chroma is 3.5 or more[2], or if the A or Ap horizon that has both low value and low chroma is too thin to be a mollic or an umbric epipedon. Epipedons that have a color value after rubbing lower than 5.5, dry, and lower than 3.5, moist, are also ochric provided they are no darker than the 1C horizon and do not have as much as 0.6 percent more organic carbon than the 1C horizon. The ochric epipedon includes eluvial horizons that are at or near

---

[2] See footnote 1 in this chapter.

the surface (the E horizon and an albic horizon, which is defined later) and extends to the first underlying diagnostic illuvial horizon (defined later as an argillic, natric, or spodic horizon). If the underlying horizon is a B horizon of alteration (defined later as a cambic or oxic horizon) and there is no surface horizon that is appreciably darkened by humus, the most convenient lower limit of the ochric epipedon is the base of the plow layer or an equivalent depth in a soil that has not been plowed. Actually, the same subhorizon in an unplowed soil may be both a part of the epipedon and a part of the cambic horizon. The epipedon and the subsurface diagnostic horizons are not mutually exclusive. The ochric epipedon does not have rock structure. It does not include fresh sediments that are finely stratified.

Plaggen epipedon (Ger. *plaggen*, sod)

The plaggen epipedon is a manmade surface layer 50 cm (20 in.) or more thick that has been produced by long-continued manuring.

The color of a plaggen epipedon and its organic-carbon content depend on sources of the materials used for bedding.

The plaggen epipedon can be identified by several means. Commonly it contains artifacts, such as bits of brick and pottery, throughout its depth. Chunks of diverse materials, such as black sand and light gray sand, as large as the size held by a spade may be present. The plaggen epipedon normally shows spade marks throughout its depth and also remnants of thin stratified beds of sand that probably were produced on the surface by beating rains and later were buried by spading.

Umbric epipedon (L. *umbra*, shade, hence dark)

Requirements of the umbric epipedon are comparable to those of the mollic epipedon in color, organic-carbon and phosphorus content, consistence, structure, n value, and thickness. The umbric epipedon includes those thick, dark-colored surface horizons, that have base saturation of <50 percent (by NH_4OAc). It should be noted that the restriction against a hard or very hard and massive epipedon when dry is applied only to those epipedons that become dry. If the epipedon is always moist, there is no restriction on its consistence or structure when dry. It should also be noted that some plaggen epipedons, defined later, meet all these requirements but also have evidences of slow addition of materials under cultivation. The umbric epipedon does not have the artifacts, spade marks, and raised surfaces that are evidences of slow additions in the plaggen epipedon.

Diagnostic Subsurface Horizons

The horizons discussed in this section form below the surface of the soil, although in some places they form immediately below a layer of leaf litter. They may be exposed at the surface by truncation of the soil. Some of these horizons are generally considered to be B horizons; some are considered B horizons by many but not all pedologists; others are generally considered to be parts of the A horizon.

Agric horizon

The agric horizon (L. *ager*, field) is an illuvial horizon
formed under cultivation that contains significant amounts
of illuvial silt, clay, and humus. After long-continued cultivation, changes in the horizon immediately below the plow
layer become apparent and cannot be ignored in classifying
the soil. The worm channels, root channels, or ped surfaces
become coated with a dark-colored mixture of organic matter, silt, and clay. The accumulation on the sides of wormholes becomes thick and eventually can fill them. If worms
are scarce, the accumulation may take the form of thick
lamellae that may range in thickness from a few millimeters
to about 1 cm. The coatings on the sides of wormholes and
lamellae always have lower color value and chroma than the
soil matrix.

The agric horizon has somewhat different forms in different
climates if there are differences in soil fauna. In a humid
temperate climate, where soils have what is later defined as
an udic moisture regime and a mesic temperature regime,
earthworms can become abundant. If there are earthworm
holes that, with their coatings, constitute 5 percent or more
of the volume and if the coatings are 2 mm or more thick
and have color value of 4 or less and chroma of 2 or less,
moist, the horizon is considered an agric horizon. After
long-continued cultivation, the content of organic matter is
not likely to be high, but the carbon-nitrogen ratio in the
agric horizon is low, usually <8. The pH of the agric horizon is close to neutrality, 6 to 6.5.

In a Mediterranean climate, where soils have what is later
defined as a xeric soil moisture regime, earthworms are less
common and the illuvial materials accumulate as lamellae
directly below the Ap horizon. If the lamellae are 5 mm or
more thick, have color value of 4 or less, moist, and chroma
of 2 or less, and constitute 5 percent or more by volume of
a horizon that is 10 cm or more thick, the horizon is considered an agric horizon.

Albic horizon

The albic (L. *albus*, white) horizon is one from which clay
and free iron oxides have been removed or in which the
oxides have been segregated to the extent that the color of
the horizon is determined by the color of the primary sand
and silt particles rather than by the coatings on these particles. An albic horizon may be at the surface of the mineral
soil; it may lie just above an argillic or a spodic horizon; it
may lie between a spodic horizon and either a fragipan or
an argillic horizon; or it may lie between an argillic horizon
and a fragipan or between a cambic horizon and an argillic
horizon, natric horizon, or fragipan. It is usually underlain
by a spodic, natric, or argillic horizon, a fragipan, or a relatively impervious layer that can produce a perched water
table and either stagnant or moving water.

Deep deposits of pure white sand can be formed by wind or
wave action. Although these deposits have the apparent
morphology of an albic horizon, they are in fact a parent
material. The white sand in such a deposit does not overlie

a B horizon or any other soil horizon except, in some places, a buried soil.

An albic horizon, therefore,is defined as a surface or a lower horizon that has such thin or discontinuous coatings on the sand or silt particles that the hue and chroma of the horizon are determined chiefly by the color of the sand and silt particles.

The color value, moist, of an albic horizon is 4 or more, or the value, dry, is 5 or more, or both. If the value, dry, is 7 or more, or the value, moist, is 6 or more, the chroma is 3 or less either dry or moist. If the value, dry, is 5 or 6, or the value, moist, is 4 or 5, the chroma is closer to 2 than to 3 either dry or moist. If parent materials have a hue of 5YR or redder, a chroma, moist, of 3 is permitted in the albic horizon if the chroma is due to the color of uncoated silt or sand grains. Under an albic horizon there is usually a B horizon that is an argillic or a spodic horizon, but in some few sandy soils the underlying horizon is too weakly developed to meet the levels of accumulation required for those horizons.

Argillic horizon

In summary, we can say that an argillic horizon is one that contains illuvial layer-lattice clays. This horizon forms below an eluvial horizon, but it may be at the surface if a soil has been partially truncated. It has the following properties that can be used for identification:

1. If an eluvial horizon remains and if there is no lithologic discontinuity between it and the argillic horizon, the argillic horizon contains more total clay and more fine clay than the eluvial horizon, as follows. The increases in clay are reached within a vertical distance of 30 cm or less.
   a. If any part of the eluvial horizon has <15 percent total clay in the fine-earth fraction (<2 mm), the argillic horizon must contain at least 3 percent more clay (13 percent versus 10 percent, for example). The ratio of fine clay to total clay normally is greater in the argillic horizon than in the overlying eluvial horizons or the underlying horizon by about one-third or more.
   b. If the eluvial horizon has >15 percent and <40 percent total clay in the fine-earth fraction, the ratio of clay in the argillic horizon to that in the eluvial horizon must be 1.2 or more. The ratio of fine clay to total clay in the argillic horizon is normally greater than in the eluvial horizon by about one-third or more.
   c. If the eluvial horizon has >40 percent total clay in the fine-earth fraction, the argillic horizon must contain at least 8 percent more clay or, if the total clay content exceeds 60 percent, 8 percent more fine clay (50 percent versus 42 percent, for example).

2. An argillic horizon should be at least one-tenth as thick as the sum of the thickness of all overlying horizons, or it should be 15 cm or more thick if the eluvial and illuvial horizons together are more than 1.5 m thick. If the argillic horizon is sand or loamy sand, it should be at least 15 cm thick. If it is composed entirely of lamellae, lamellae ≥ 1 cm

thick should have a combined thickness of at least 15 cm. If the argillic horizon is loamy or clayey, it should be at least 7.5 cm thick.

3. In structureless soils, the argillic horizon has oriented clay bridging the sand grains and also in some pores.

4. If peds are present, an argillic horizon should meet one of the following requirements:
  a. Have clay skins on some of both the vertical and horizontal ped surfaces and in the fine pores or have oriented clay in 1 percent or more of the cross section;
  b. Meet requirements 1 and 2 and also have a broken or irregular upper boundary and some clay skins in the lowest part of the horizon;
  c. If the horizon is clayey, if the clay is kaolinitic, and if the surface horizon has >40 percent clay, have some clay skins on peds and in pores in the lower part of the horizon that has blocky or prismatic structure; or
  d. If the illuvial horizon is clayey with 2-to-1 lattice clays, an argillic horizon does not need to have clay skins if there are uncoated grains of sand or silt in the overlying horizon and evidences of pressure caused by swelling or if the ratio of fine to total clay in the horizon is greater by at least one-third than the ratio in the overlying or the underlying horizon or if it has >8 percent more fine clay. The evidences of pressure may be occasional slickensides or wavy horizon boundaries in the illuvial horizon.

5. If a soil has a lithologic discontinuity between the eluvial horizon and the argillic horizon or if only a plow layer overlies the argillic horizon, the argillic horizon needs to have clay skins in only some part, either in some fine pores or, if peds exist, on some vertical and horizontal ped surfaces. Either thin sections should show that some part of the horizon has about 1 percent or more of oriented clay bodies or the ratio of fine clay to total clay should be greater than in the overlying or the underlying horizon.

Calcic horizon and k horizon

The calcic horizon is a horizon of accumulation of calcium carbonate or of calcium and magnesium carbonate. The accumulation may be in the C horizon, but it may also be in a variety of other horizons such as a mollic epipedon, an argillic or a natric horizon, or a duripan.

The calcic horizon has two forms. In one, the underlying materials have less carbonate than the calcic horizon. This form of calcic horizon includes horizons of secondary carbonate enrichment that are 15 cm (6 in.) or more thick, have a carbonate content equivalent to ≥ 15 percent $CaCO_3$, and have a $CaCO_3$ equivalent at least 5 percent greater than the C horizon. In the other form, the calcic horizon is 15 cm or more thick, has a $CaCO_3$ equivalent ≥ 15 percent, and contains ≥ 5 percent, by volume, of identifiable secondary carbonates as pendants on pebbles, concretions, or soft powdery forms. If this calcic horizon rests on limestone, marl, or other very highly calcareous materials (≥ 40 percent $CaCO_3$

equivalent), the percentage of carbonates need not decrease with depth.

If the particle-size class is sandy, sandy-skeletal, coarse-loamy, or loamy-skeletal with less than 18 percent clay, the 15 percent requirement for $CaCO_3$ equivalent is waived. But to qualify as a calcic horizon, the horizon must have at least 5 percent (by volume) more soft powdery secondary $CaCO_3$ than an underlying horizon, and the calcic horizon must be at least 15 cm thick.

If a horizon enriched with secondary carbonate is indurated or cemented to the degree that dry fragments do not slake in water, it is considered to be a petrocalcic horizon, which is discussed later. Air-dry fragments of a calcic horizon will slake in water. Pendants below rocks and concretions normally do not slake, but these are not connected, and the soil material between the concretions will slake.

Cambic horizon

In summary, the cambic horizon is an altered horizon that does not have the dark color, organic-matter content, and structure that are definitive of a histic, a mollic, or an umbric epipedon, and it has

1. Texture that is very fine sand, loamy very fine sand, or finer in the fine-earth (<2 mm) fraction;

2. Soil structure or absence of rock structure in at least half the volume;

3. Minerals that consist of (a) enough amorphous or 2:1 lattice clay to give a cation-exchange capacity (by NH_4OAc) of more than 16 cmol(+) kg^{-1} clay or (b) ≥10 percent weatherable minerals;

4. Evidence of alteration in one of the following forms:
   a. Gray colors as defined for an aquic moisture regime, defined later, or artificial drainage, and one or more of the following properties:
      (1) A regular decrease in the amount of organic carbon with depth and a content of <0.2 percent organic carbon at a depth of 1.25 m below the surface or immediately above a sandy-skeletal substratum that is at a depth of <1.25 m;
      (2) Cracks that open and close in most years and are 1 cm or more wide at a depth 50 cm below the surface;
      (3) Permafrost at some depth;
      (4) A histic epipedon consisting of mineral soil materials or a mollic or umbric epipedon;
   b. Stronger chroma, redder hue, or higher clay content than the underlying horizon;
   c. Evidences of removal of carbonates. Particularly, the cambic horizon has less carbonate than the underlying k horizon. If all coarse fragments in the k horizon are completely coated with lime, some in the cambic horizon are partly free of coatings. If coarse fragments in the k horizon are coated only on the under side, those in the cambic horizon should be free of coatings;

d. If carbonates are absent in the parent material and in the dust that falls on the soil, the required evidence of alteration is satisfied by the presence of soil structure and the absence of rock structure if the moisture regime is not aquic or the chroma is higher than those listed:

(1). if there is mottling, the chroma is 2 or less;

(2). if there is no mottling and the value is less than 4, the chroma is less than 1; if the value is 4 or more, the chroma is 1 or less;

(3). the hue is no bluer than 10Y if the hue changes on exposure to air (a hue bluer than 10Y that does not change on exposure is not diagnostic);

5. Properties that do not meet the requirements of an argillic, kandic, or spodic horizon;

6. No cementation or induration and no brittle consistence when moist; and

7. Enough thickness that its base is at least 25 cm (10 in.) below the soil surface unless the soil temperature regime is cryic or pergelic.

Duripan

The duripan (L. *durus*, hard, plus pan; meaning hardpan) is a subsurface horizon that is cemented by silica to the degree that fragments from the air-dry horizon do not slake during prolonged soaking in water or in HCl.

In summary, the duripan is a silica-cemented subsurface horizon in which

1. Cementation is strong enough that dry fragments from some subhorizon do not slake in water, even during prolonged wetting;

2. Coatings of silica, insoluble in 1 M HCl even during prolonged soaking but soluble in hot concentrated KOH or in alternating acid and alkali, are present in some pores and on some structural faces; or some durinodes are present; and

3. Cementation is not destroyed by soaking in acid in more than half of any laminar capping that may be present or in some other continuous or imbricated subhorizon. Cementation in such layers is completely destroyed by hot concentrated KOH, either by a single treatment or by alternating with acid.

4. If fractured, the average lateral distance between fracture points is 10 cm or more.

Fragipan

A fragipan (modified from L. *fragilis*, brittle, and pan; meaning brittle pan) is a loamy or uncommonly a sandy subsurface horizon that may but does not necessarily underlie a cambic, spodic, argillic, or albic horizon. It has a very low content of organic matter, has high bulk density relative to the horizons above it, and is seemingly cemented when dry, having then hard or very hard consistence. When moist, a fragipan has moderate or weak brittleness, which is the tendency for a ped or clod to rupture suddenly when pressure is applied rather than to undergo slow deformation. A dry fragment slakes or fractures when placed in water. A fragipan is usually mottled, is slowly or very slowly permeable to water, and has few or many bleached, roughly vertical planes that are faces of coarse or very coarse polyhedrons or prisms.

<u>Identification</u>

There is no known laboratory procedure for identifying a sample of a fragipan. Identification is primarily a field problem. A combination of clues must be used because there is no single unique property of fragipans. First, position is important. A fragipan lies below an eluvial horizon unless the soil has been truncated, but it is not necessarily immediately below. If the soil has been truncated, the pan can be traced up slope until it lies under an eluvial horizon.

Second, if there is an argillic or a cambic horizon above a fragipan, there is commonly an E' horizon between the fragipan and the overlying horizon. The E' horizon is marked by uncoated grains of sand and silt. This horizon seems related to water that either stands above the pan or moves laterally along its surface.

Third, if the pan is not saturated for long periods, some or all pedons normally have bleached vertical streaks that form a roughly polygonal pattern on a horizontal plane. The bleached streaks are bounded by strong brown or reddish brown streaks where iron and manganese have accumulated. If the pan is saturated for long periods or if the texture is sandy, the polygonal color pattern may be absent.

Fourth, if the moisture content is near the wilting point, the matrix between the streaks is very firm. If it is near field capacity, the matrix is brittle. The brittle matrix should constitute 60 percent or more of the volume of some subhorizon.

Fifth, fine feeder roots are virtually absent in the brittle parts of a fragipan. If brittleness is so weakly expressed that fine feeder roots are present throughout the horizon, the horizon should not be considered a fragipan. It should be noted, however, that some trees have tap roots that extend through a well-expressed fragipan, but this is the exception rather than the rule. It is characteristic of fragipans that few or many roots may be present in the bleached vertical streaks and that few or no fine roots are present in the brittle matrix between the bleached streaks. The fine roots should not be present at intervals of <10 cm except within bleached vertical streaks, and the mean horizontal dimensions of the brittle matrix should be at least 10 cm.

Sixth, texture of the fine-earth fraction of a fragipan is finer than fine sand, and the percentage of clay is generally <35; in most soils appreciably less. The texture normally is loamy, that is, silt loam, loam, or sandy loam.

Seventh, an air-dry fragment about the size of a fist slakes or fractures when placed in water.

Gypsic horizon

The gypsic horizon is a noncemented or weakly cemented horizon of enrichment with secondary sulfates that is 15 cm or more thick, has at least 5 percent more gypsum than the C horizon or the underlying stratum, and in which the product of the thickness in centimeters and the percentage of gypsum is ≥ 150. Thus, a horizon 30 cm thick that has 5 percent gypsum qualifies if gypsum is absent in the underlying horizon. A layer 30 cm thick that has 6 percent gypsum qualifies if the gypsum content of the underlying horizon is not more than 1 percent. Cementation is weak enough that a dry fragment slakes in water.

The percentage of gypsum can be calculated by multiplying the milliequivalents of gypsum per 100 g soil by the milliequivalent weight of gypsum, which is 0.086.

Kandic horizon

The kandic horizon:

1. Is a vertically continuous subsurface horizon and has, starting at the point where the clay increase requirements are met, a CEC of <16 cmol(+) kg^{-1} clay (by 1 M NH_4OAc pH 7) and an ECEC <12 cmol(+) kg^{-1} clay (sum of bases extracted with 1 M NH_4OAc pH 7 plus 1 M KCl-extractable Al) in at least the major part of the horizon.

2. Has a thickness of at least 30 cm, or if a lithic, paralithic, or petroferric contact occurs within 50 cm of the soil surface, then the thickness of the kandic horizon is at least 60 percent of the vertical distance between 18 cm and the contact but at least 15 cm thick.

3. Has a texture of loamy very fine sand or finer.

4. Underlies a coarser textured surface horizon. The minimum thickness of the surface horizon is 18 cm after mixing, or 5 cm if the textural transition to the kandic horizon is abrupt and if there is no lithic, paralithic, or petroferric contact within 50 cm.

5. Has more total clay than the overlying coarser textured surface horizon and the increased clay content is reached within a vertical distance of 15 cm or less as follows:
   a. If the surface horizon as defined above has less than 20 percent total clay, the kandic horizon begins where some subhorizon contains at least 4 percent more clay absolute than the overlying horizon.
   b. If the surface horizon as defined above has 20 to 40 percent total clay the kandic horizon begins where some subhorizon has at least 1.2 times more clay than the overlying horizon.
   c. If the surface horizon as defined above has more than 40 percent total clay, the kandic horizon begins where some subhorizon has at least 8 percent more clay absolute than the overlying horizon.

Natric horizon

The natric horizon (NL. *natrium*, sodium; implying presence of sodium) is a special kind of argillic horizon. It has, in addition to the properties of the argillic horizon:

1. Either
    a. Prisms or, more commonly, columns in some part, usually the upper part, that may or may not break to blocks; or
    b. Rarely, blocky structure and tongues of an eluvial horizon, in which there are uncoated silt or sand grains, extending more than 2.5 cm into the horizon; <u>and</u>

2. Either
    a. The SAR^3 is ≥ 13 (or 15 percent or more saturation with exchangeable sodium) in some subhorizon within 40 cm of the upper boundary; or
    b. More exchangeable magnesium plus sodium than calcium plus exchange acidity (at pH 8.2) in some subhorizon within 40 cm of the upper boundary if the SAR is ≥ 13 (or ESP ≥ 15) in some horizon within 2 m of the surface.

Oxic horizon

The oxic horizon is a horizon beginning at 18 cm or more below the soil surface that:

1. Is at least 30 cm thick;

2. Has a particle-size of sandy loam or finer in the fine earth fraction;

3. Has a fine-earth fraction (<2 mm) that has an apparent ECEC (NH_4OAc bases plus 1 M KCl extractable Al) equal or less than 12 cmol(+) kg^{-1} clay and has an apparent CEC pH7 (NH_4OAc CEC) equal to or less than 16 cmol(+) kg^{-1} clay (measured clay or 3 x 1500 kPa water, whichever is greater but less than 100);

4. Does not have as much as 10 percent weatherable minerals in the 50-200 micrometer fraction;

5. Has a diffuse upper particle-size boundary (i.e., <1.2 times clay content increase within a vertical distance of 15 cm if the surface horizon contains 20-40 percent clay; less than 4 percent absolute clay content increase if the surface

---

[3]The percentage of exchangeable sodium (ESP) is used in the definition of the natric horizon and in a number of the taxa. Since this text was written, the U.S. Salinity Laboratory (personal communication from C. A. Bower) has revised its definition of sodic (alkali) soils and the method for measuring the sodium adsorption ratio (SAR) as follows: SAR is measured by the normal method if the conductivity (EC) of the saturation extract is <20 dS m^{-1} at $25^{\circ}C$. If the conductivity is ≥ 20 dS m^{-1} and SAR is >10, SAR is determined on a sample that has been leached with distilled water until EC of the leachate decreases to about 4 dS m^{-1} but not to <4. ESP of ≥ 15 is replaced by SAR of ≥ 13 if EC is large enough to require a correction for soluble salts in calculating ESP. If EC is low enough (≤ 4) that no correction is needed for soluble salts, ESP is determined directly from the replaced cations.

contains $\leq$ 20 percent clay; <8 percent absolute if the surface contains $\geq$40 percent clay;

6. Does not have andic soil properties (see section on "andic soil properties later in this chapter);

7. Has less than 5 percent by volume that shows rock structure unless the lithorelicts containing weatherable minerals are coated with sesquioxides.

Petrocalcic horizon

The petrocalcic horizon is a continuous, cemented or indurated calcic horizon that is cemented by calcium carbonate or in some places by calcium and some magnesium carbonate. Accessory silica may be present. The petrocalcic horizon is continuously cemented throughout the pedon to the degree that dry fragments do not slake in water. It cannot be penetrated by spade or auger when dry. It is massive or platy, very hard or extremely hard when dry, and very firm or extremely firm when moist. Noncapillary pores are filled, and the petrocalcic horizon is a barrier to roots. Hydraulic conductivity is moderately slow to very slow. The horizon is usually much more than 10 cm (4 in.) thick.

A laminar capping commonly is present but is not required.

If a laminar horizon rests on bedrock, it is considered a petrocalcic horizon if it is 2.5 cm of more thick and the product of the thickness in centimeters multiplied by the percentage of $CaCO_3$ equivalent is 200 or more.

Petrogypsic horizon

The petrogypsic horizon is a gypsic horizon that is strongly enough cemented with gypsum that dry fragments do not slake in water and that roots cannot enter. The gypsum content commonly is far greater than the minimum requirements for the gypsic horizon and usually exceeds 60 percent. Petrogypsic horizons are restricted to arid climates and to parent materials that are rich in gypsum.

Placic horizon

The placic horizon (Gr. base of *plas*, flat stone; meaning a thin cemented pan) is a thin, black to dark reddish pan cemented by iron, by iron and manganese, or by an iron-organic matter complex. Its thickness ranges generally from 2 mm to 10 mm. Rarely, it is as thin as 1 mm or as thick as 20 to 40 mm in spots. It may be, but not necessarily, associated with stratification in parent materials. It is in the solum, roughly parallel to the soil surface, and is commonly within the upper 50 cm of the mineral soil. It has a pronounced wavy or even convolute form. It normally occurs as a single pan, not as multiple sheets one underlying another, but in places it may be bifurcated. It is a barrier to water and roots.

An iron-cemented pan is strong brown to dark reddish brown. A pan cemented by iron and manganese or by iron-organic matter complexes is black or reddish black. A single pan may contain two or more layers cemented by different agents. Iron-organic matter cements commonly are present in the upper part of the pan.

Identification is seldom difficult. The hard brittle pan differs so much from the material in which it occurs and is so close to the surface of the mineral soil material that it is obvious unless its thickness is minimal. A few analyses of placic horizons show that organic carbon is present in amounts ranging from 1 percent to 10 percent or more. The presence of organic carbon as well as the shape and position of the pan distinguish the placic horizon from the ironstone sheet that may form where water hangs or moves laterally at a lithologic discontinuity.

Salic horizon

A salic horizon is a horizon 15 cm or more thick that contains a secondary enrichment of salts more soluble in cold water than gypsum. It contains at least 2 percent salt, and the product of its thickness in centimeters and salt percentage by weight is 60 or more. Thus, a horizon 20 cm thick would need to contain 3 percent salt to qualify as a salic horizon and a horizon 30 cm thick would need 2 percent.

Sombric horizon

The sombric horizon is a subsurface horizon of mineral soils formed under free drainage. It contains illuvial humus that is neither associated with aluminum, as is the humus in the spodic horizon, nor dispersed by sodium, as is common in the natric horizon. Consequently, the sombric horizon does not have the high cation-exchange capacity of a spodic horizon relative to clay, and it does not have the high base saturation of a natric horizon. The sombric horizon does not underlie an albic horizon.

Sombric horizons are thought to be restricted to the cool moist soils of the high plateaus and mountains in tropical or subtropical regions. Because of the annual leaching, base saturation is low, <50 percent by NH_4OAc.

The sombric horizon has lower color value or chroma, or both, than the overlying horizon and commonly, but not necessarily, contains more organic matter than the overlying horizon. It may have formed in an argillic, a cambic, or, possibly, an oxic horizon. If peds are present, the dark colors are most pronounced on ped surfaces.

A sombric horizon is easily confused in the field with a buried A horizon. It can be distinguished from some buried epipedons by lateral tracing. In thin sections, the organic matter of a sombric horizon appears more concentrated on peds and in pores than uniformly dispersed through the matrix.

Spodic horizon

A spodic horizon is normally a subsurface horizon that underlies an O, A, Ap, or E horizon. It may, however, meet the definition of an umbric epipedon. A spodic horizon has the morphological or the chemical and physical characteristics that are listed next, and its hue and chroma remain constant with increasing depth or the subhorizon that has the reddest hue or the highest chroma is near the top of the horizon. The color changes within 50 cm from the top of the horizon[4]. If the soil temperature regime is frigid or warmer, some part of the spodic horizon must meet one or more of the following requirements below a depth of 12.5 cm or below any Ap horizon that is present. If the soil temperature regime is cryic or pergelic, there is no requirement for depth. In addition, the spodic horizon must meet one or more of the following requirements:

1. Have a subhorizon > 2.5 cm thick that is continuously cemented by some combination of organic matter with iron or aluminum or with both;

2. Have a particle-size class that is sandy or coarse-loamy, and sand grains are covered with cracked coatings or there are distinct dark pellets of coarse-silt size or larger, or both; or

3. Have one or more subhorizons in which
   a. If there is 0.1 percent or more extractable iron, the ratio of iron plus aluminum (elemental) extractable by pyrophosphate at pH 10 to percentage of clay is ≥ 0.2 (percentage of pyrophosphate-extractable Fe + Al at pH 10/clay percentage ≥ 0.2) or if there is <0.1 percent extractable iron, the ratio of aluminum plus carbon extractable by pyrophosphate at pH 10 to percentage clay is ≥ 0.2.
   b. The sum of pyrophosphate-extractable iron plus aluminum is half or more of the extractable sum of dithionite-citrate extractable iron plus aluminum (percentage of pyrophosphate-extractable Fe + Al/percentage of dithionite-citrate extractable Fe + Al ≥ 0.5); and
   c. The combined index of accumulation of amorphous material must be 65 or more. The index for each subhorizon is calculated by subtracting half of the clay percentage from CEC at pH 8.2 and multiplying the remainder by the thickness of the subhorizon in centimeters. The results for all subhorizons are then added and the total must be 65 or more.

Sulfuric horizon

The sulfuric (L. *sulfur*) horizon is composed either of mineral or organic soil material that has both a pH <3.5 (1:1 in water) and jarosite mottles (the color of fresh straw that has a hue of 2.5Y or yellower and chroma of 6 or more).

A sulfuric horizon forms as a result of artificial drainage and oxidation of sulfide-rich mineral or organic materials. Such a horizon is highly toxic to plants and virtually free of living roots.

---

[4] A thin black horizon that has color value of 2 or less may overlie this horizon.

Other Diagnostic Soil Characteristics

Abrupt textural change

An abrupt textural change is a change from an ochric epipedon or an albic horizon to an argillic horizon. There is, in the zone of contact, a very appreciable increase in clay content within a very short distance in depth. If the clay content of the ochric epipedon or the albic horizon is <20 percent, the clay content should double within a distance in depth of 7.5 cm or less. If the clay content exceeds 20 percent, the increase in clay content should be at least 20 percent of the fine-earth fraction, for example, from 22 percent to 42 percent, within a distance of 7.5 cm in depth, and the clay content in some part of the argillic horizon should be at least double that of the horizon above.

A transitional horizon normally is not present or is too thin to be sampled. In some soils, however, there may be tonguing or interfingering of albic materials, which are defined later, in parts of the argillic horizon. The horizon boundary in such a soil is irregular or even discontinuous. The sampling of such a mixture as a single horizon might create the impression of a relatively thick transitional horizon, even though the thickness of the actual transition at the contact may be only 1 mm or so.

Amorphous material dominant in the exchange complex

Amorphous material, as the term is used here, is colloidal material that includes allophane and has all or most of the properties of allophane. The term is more inclusive, however, than allophane as it is defined by some workers. Amorphous material, as used here, is generally amorphous under X-ray analysis, but enough crystalline materials may be present, especially in mixtures, to cause small and disordered peaks. The amorphous material is associated with organic matter, but it contains aluminum, and it never has more than traces of aluminum that can be extracted with KCl. Consequently, if the base saturation is low, that is, <35 percent, the amorphous material has a permanent charge of less than 10 cmol(+) kg^{-1}. It has high exchange capacity, however, in a system buffered at pH 7, and very high exchange capacity at pH 8.2. The exchange capacity is clearly pH induced. The amorphous material also has high anion-exchange capacity. It has an enormous surface area and retains much water against 1500 kPa tension, commonly 50 to 100 percent or more. It cannot be dispersed readily in hexametaphosphate.

If amorphous material dominates an exchange complex, we find that the following conditions are satisfied:

1. The exchange capacity of the clay at pH 8.2 is >150 cmol(+) kg^{-1} measured clay, and commonly is >500 cmol(+) kg^{-1}. The high value is, in part, the result of the poor dispersion.

2. If there is enough clay to have a 1500 kPa water content of 20 percent or more, the pH of a suspension of 1 g soil in 50 ml 1 M NaF is >9.4 after 2 minutes.

3. The ratio of 1500 kPa water content to measured clay is more than 1.0.

4. The amount of organic carbon exceeds 0.6 percent.

5. Differential thermal analysis shows a low-temperature endotherm.

6. The bulk density of the fine-earth fraction is <0.85 g per cubic centimeter at 33 kPa tension.

Andic soil properties

Andic soil properties are tentatively defined as soil materials meeting one or more of the following three requirements:
  1.a. Acid oxalate extractable aluminum plus 1/2 acid oxalate extractable iron is 2.0 percent or more in the <2 mm fraction;
  b. Bulk density of the <2 mm fraction, measured at 33 kPa water retention, is $\leq$0.90 g cm^{-3}; and
  c. Phosphate retention is more than 85 percent; or
  2.a. More than 60 percent by volume of the whole soil is volcaniclastic material coarser than 2 mm; and
  b. Acid oxalate extractable aluminum plus 1/2 acid oxalate extractable iron is 0.40 percent or more in the <2 mm fraction; or
  3. The 0.02 to 2.0 mm fraction is at least 30 percent of the <2 mm fraction and meets one of the following:
  a. If the <2 mm fraction has acid oxalate extractable aluminum plus 1/2 acid oxalate extractable iron of 0.40 percent, there is at least 30 percent volcanic glass in the 0.02 to 2.0 mm fraction;
  b. If the <2 mm fraction has acid oxalate extractable aluminum plus 1/2 acid oxalate extractable iron of 2.0 percent or more, there is at least 5 percent volcanic glass in the 0.02 to 2.0 mm fraction; or
  c. If the <2 mm fraction has acid oxalate extractable aluminum plus 1/2 acid oxalate extractable iron of between 0.40 percent and 2.0 percent, there is a proportional content of volcanic glass in the 0.02 to 2.0 mm fraction between 5 and 30 percent.

Coefficient of linear extensibility, COLE

This coefficient is the ratio of the difference between the moist length and the dry length of a clod to its dry length. It is $(Lm - Ld)/Ld$, where Lm is the length at 33 kPa tension and Ld is the length when dry. It can be calculated from the difference in bulk density of the clod when moist and when dry. COLE can be estimated from shrinkage of a sample that has been packed at field capacity into a mold and then dried.

Durinodes

Durinodes (L. *durus*, hard; *nodus*, knot) are weakly cemented to indurated nodules. The cement is SiO_2, presumably opal and microcrystalline forms of silica. It breaks down in hot concentrated KOH after treatment with HCl to remove carbonates but does not break down with concentrated HCl alone. Dry durinodes do not slake appreciably in water, but prolonged soaking can result in spalling of very thin platelets and some slaking. The durinodes are firm or very firm; they are brittle when wet, both before and after treatment with acid; and they are disconnected and they range upward in size from a diameter of about 1 cm. Most durinodes are roughly concentric when viewed in cross section, and concentric stringers of opal may be visible under a hand lens.

Gilgai

Gilgai is the microrelief that is typical of clayey soils that have a high coefficient of expansion with changes in moisture content and that also have distinct seasonal changes in moisture content. The microrelief consists of either a succession of enclosed microbasins and microknolls in nearly level areas or of microvalleys and microridges that run up and down the slope. The height of the microridges commonly ranges from a few centimeters to 1 m. Rarely does the height approach 2 m.

Lithic contact

A lithic contact is a boundary between soil and coherent underlying material. Except in Ruptic-Lithic subgroups the underlying material must be continuous within the limits of a pedon except for cracks produced in place without significant displacement of the pieces. Cracks should be few, and their average horizontal spacing should be 10 cm or more. The underlying material must be sufficiently coherent when moist to make hand digging with a spade impractical, although it may be chipped or scraped with a spade. If it is a single mineral, it must have a hardness by Mohs scale of 3 or more. If it is not a single mineral, chunks of gravel size that can be broken out must not disperse during shaking for 15 hours in water or in sodium hexametaphosphate solution. The underlying material considered here does not include diagnostic soil horizons such as duripan or a petrocalcic horizon.

A lithic contact is diagnostic at the subgroup level if it is within 125 cm of the soil surface of Oxisols and within 50 cm of the soil surface of all other mineral soils.

Mottles that have chroma of 2 or less

It refers to colors in a horizon in which parts have chroma of 2 or less, moist, and value, moist, of 4 or more, whether or not that part is dominant in volume or whether or not it is a continuous phase surrounding spots of higher chroma. If either the minor or major part of a horizon has chroma of 1 to 2 and value, moist, or 4 or more and there are spots of higher chroma, the part that has the lower chroma is included in the meaning of "mottles that have chroma of 2 or less." The part is excluded from the meaning if all the horizon has chroma of 2 or less or if no part of the horizon has chroma as low as 2.

The phrase also means that the horizon that has such mottles is saturated with water at some period of the year or the soil is artificially drained. It is also implicit in the meaning that the temperature of the horizon is above the biologic zero, which is about 5°C (41°F), during at least a part of the time that the horizon is saturated.

n value

The *n* value (Pons and Zonneveld 1965) refers to the relation between the percentage of water under field conditions and the percentages of inorganic clay and of humus. The *n* value is helpful in predicting whether the soil may be grazed by livestock or will support other loads and the degree of subsidence that would occur after drainage. The *n* value can be calculated for mineral soil materials that are not thixotropic by the formula:

$$n = (A - 0.2R)/(L + 3H)$$

A is the percentage of water in the soil in field condition, calculated on a dry-soil basis; *R* is the percentage of silt plus sand; *L* is the percentage of clay; and *H* is the percentage of organic matter (organic carbon x 1.724).

Few data are available in the United States for calculations of the *n* value, but the critical *n* value of 0.7 can be approximated closely in the field by a simple test of squeezing the soil in the hand. If the soil flows with difficulty between the fingers, the *n* value is between 0.7 and 1.0. If the soil flows easily between the fingers, the *n* value is 1 or more.

Organic soil materials

Organic soil materials either

1. Are saturated with water for long periods or are artificially drained, and have 18 percent or more organic carbon if the mineral fraction is 60 percent or more clay, 12 percent or more organic carbon if the mineral fraction has no clay, or a proportional amount of organic carbon between 12 and 18 percent if the clay content is between zero and 60 percent; or

2. Are never saturated with water for more than a few days and have 20 percent or more organic carbon.

Item 1 in this definition covers materials that have been called peat and muck. Item 2 is intended to include materials that have been called "litter" or O horizons. Not all organic soil materials accumulate under water. Leaf litter may rest on a lithic contact and yet may support a forest. The only "soil" in this situation is organic in the sense that the mineral fraction may be appreciably less than half the weight and only a small proportion of the volume of the soil.

Paralithic contact

A paralithic (lithic-like) contact is a boundary between soil and continuous coherent underlying material. It differs from a lithic contact in that the underlying material, if a single mineral, has a hardness by Mohs scale of <3. If the underlying material is not a single material, chunks of gravel size that can be broken out disperse more or less completely during 15 hours of end-over-end shaking in water or in sodium hexametaphosphate solution and, when moist, the material can be dug with difficulty with a spade.

Permafrost

Permafrost is a layer in which the temperature is perennially at or below 0°C, whether the consistence is very hard or loose. Dry permafrost has loose consistence.

Petroferric contact

A petroferric (Gr. *petra*, rock, and L. *ferrum*, iron; implying ironstone) contact is a boundary between soil and a continuous layer of indurated material in which iron is an important cement and organic matter is absent or is present only in traces. The indurated layer must be continuous within the limits of a pedon but may be fractured if the average lateral distance between fractures is $\geq$10 cm. The indurated layer is distinguished from a placic horizon and from an indurated spodic horizon (ortstein) because it contains little or no organic matter. Organic matter is present in both the other horizons.

Several features can aid in making the distinction between a lithic and a petroferric contact. First, a petroferric contact is roughly horizontal. Second, the amount of iron in the material immediately below a petroferric contact is high. The content of Fe_2O_3 normally ranges upward from 30 percent. Third, the ironstone sheets below a petroferric contact are thin. Their thickness ranges from a few centimeters to a very few meters. Sandstone, on the other hand, may be thin or very thick, may be level bedded or tilted, and may have only a small percentage of Fe_2O_3. In the Tropics the ironstone commonly is more or less vesicular.

Plinthite

Plinthite (Gr. *plinthos*, brick) is an iron-rich, humus-poor mixture of clay with quartz and other dilutents. It commonly occurs as dark red mottles, which usually are in platy, polygonal, or reticulate patterns. Plinthite changes irreversibly to an ironstone hardpan or to irregular aggregates on exposure to repeated wetting and drying, especially if it is exposed also to heat from the sun. The lower boundary of a zone in which plinthite occurs usually is diffuse or gradual, but it may be abrupt at a lithologic discontinuity.

Generally, plinthite forms in a horizon that is saturated with water at some season. The original segregation of the iron normally is in the form of soft, more or less clayey, red or dark red mottles. The mottles are not considered plinthite unless there has been enough segregation of iron to permit irreversible hardening on exposure to wetting and drying. Plinthite in the soil usually is firm or very firm when the soil moisture content is near field capacity and hard when the moisture content is below the wilting point. Plinthite does not harden irreversibly as a result of a single cycle of drying and rewetting. After a single drying, it will remoisten, and then it can be dispersed in large part by shaking in water with a dispersing agent.

In a moist soil, plinthite is soft enough that it can be cut with a spade. After irreversible hardening, it is no longer considered plinthite but is called ironstone. Indurated ironstone materials can be broken or shattered with a spade but cannot be dispersed by shaking in water with a dispersing agent.

Potential linear extensibility

This characteristic is the sum of the products, for each horizon, of the thickness of the horizon in centimeters and the COLE of the horizon.

Sequum: number and kind

A sequence of an eluvial horizon and its subjacent B horizon, if one is present, is called a sequum. An albic horizon and a spodic horizon immediately underlying it, for example, constitute a sequum. Similarly, a mollic epipedon over a cambic horizon or an argillic horizon over a k horizon also constitute a sequum. Two sequa may be present in vertical sequence in a single soil, and that sequence is called a bisequum.

Slickensides

Slickensides are polished and grooved surfaces that are produced by one mass sliding past another. Some of them occur at the base of a slip surface where a mass of soil moves downward on a relatively steep slope. Slickensides are very common in swelling clays in which there are marked changes in moisture content.

Soft powdery lime

Soft powdery lime is a phrase that is used in the definitions of a number of taxa. It refers to translocated authigenic lime, soft enough to be cut readily with a fingernail, that was precipitated in place from the soil solution rather than inherited from a soil parent material such as a calcareous loess or till. It should be present in a significant enough accumulation to constitute a k horizon.

To be identifiable, soft powdery lime must have some relation to the soil structure or fabric. It may disrupt the fabric to form spheroidal aggregates, or white eyes, that are soft and powdery when dry. Or the lime may be present as soft coatings in pores or on structural faces. If present as coatings, it covers a significant part of the surface. Commonly, it coats the whole surface to a thickness of 1 to 5 mm or more. But only part of a surface may be coated if little lime is present in the soil. The coatings should be thick enough to be visible when moist and should cover a continuous area large enough to be more than filaments. The pseudomycelia commonly seen in a dry calcareous horizon do not come within the meaning of soft powdery lime. Pseudomycelia are soft powdery filaments on structural faces, commonly branching, but they may come and go with the seasons and may be only lime that was precipitated in a single season by the withdrawal of stored soil moisture rather than a k horizon.

Soft coatings on hard lime concretions are also excluded from the meaning of soft powdery lime. These may be thin or thick, and they may be the result of either current accumulation or removal of lime. That is, the concretion may be growing or may be undergoing dissolution, and either process can produce a soft coating.

Soil moisture regimes

The soil moisture regime, as the term is used here, refers to the presence or absence either of ground water or of water held at a tension <1500 kPa in the soil or in specific horizons by periods of the year. Water held at a tension of 1500 kPa or more is not available to keep most mesophytic plants alive. The availability of water also is affected by dissolved salts. A soil may be saturated with water that is too salty to be available to most plants, but it would seem better to call such a soil salty rather than dry. Consequently, we consider a horizon to be dry when the moisture tension is 1500 kPa or more. If water is held at a tension of <1500 kPa but more than zero, we consider the horizon to be moist. A soil may be continuously moist in some or all horizons throughout the year or for some part of the year. It may be moist in winter and dry in summer or the reverse. In the northern hemisphere, summer refers to the months of June, July, and August, and winter means December, January, and February. A soil or a horizon is considered to be saturated with water when water stands in an unlined borehole close enough to the soil surface or to the horizon in question that the capillary fringe[5] reaches the surface or the top of the horizon.

<u>Soil moisture control section</u>

The intent in defining the soil moisture control section is to facilitate estimation of soil moisture regimes from climatic data. The upper boundary of this control section is the depth to which a dry (tension >1500 kPa but not air dry) soil will be moistened by 2.5 cm (1 in.) of water within 24 hours. The lower boundary is the depth to which a dry soil will be moistened by 7.5 cm (3 in.) of water within 48 hours. These depths are exclusive of the depth of moistening along any cracks or animal burrows that are open to the surface.

If 7.5 cm of water moistens the soil to a lithic, petroferric, or paralithic contact or to a petrocalcic horizon or a duripan, the upper boundary of the rock or of the cemented horizon is the lower boundary of the soil moisture control section. If 2.5 cm of water moistens the soil to one of these contacts or horizons, the soil moisture control section is the lithic contact itself, the paralithic contact, or the upper boundary

---

[5]The capillary fringe is the zone just above the water table (zero gauge pressure) that remains almost saturated (Soil Sci. Soc. Amer. Glossary, 1965, p.332).

of the cemented horizon. The control section of the latter soil is moist if the upper boundary of the rock or the cemented horizon has a thin film of water. If the upper boundary is dry, the control section is dry.

As a rough guide to the limits, the soil moisture control section lies approximately between 10 and 30 cm (4 and 12 in.) if the particle-size class is fine-loamy, coarse-silty, fine-silty, or clayey. The control section extends approximately from a depth of 20 cm to a depth of 60 cm (8 to 24 in.) if the particle-size class is coarse-loamy, and from 30 to 90 cm (12 to 35 in.) if the particle-size class is sandy.

<u>Classes of soil moisture regimes</u>

The moisture regimes are defined in terms of the ground-water level and in terms of the presence or absence of water held at a tension <1500 kPa in the moisture control section by periods of the year. It is assumed in the definitions that the soil supports whatever vegetation it is capable of supporting. In other words, it is in crops, grass, or native vegetation; it is not being fallowed to increase the amount of stored moisture, nor is it being irrigated by man. These cultural practices affect the soil moisture condition as long as they are continued.

Aquic moisture regime.--The aquic (L. *aqua*, water) moisture regime implies a reducing regime that is virtually free of dissolved oxygen because the soil is saturated by ground water or by water of the capillary fringe. An aquic regime must be a reducing one. Some soil horizons, at times, are saturated with water while dissolved oxygen is present, either because the water is moving or because the environment is unfavorable for micro-organisms, for example, if the temperature is <1°C such a regime is not considered aquic.

For differentiation in the highest categories of soils that have an aquic regime, the whole soil must be saturated. In the subgroups, only the lower horizons are saturated. The soil is considered to be saturated if water stands in an unlined borehole at such a shallow depth that the capillary fringe (see footnote 5) reaches the soil surface except in noncapillary pores. The water in the borehole is stagnant and remains colored if a dye is placed in the water. In a sandy soil, the thickness of the capillary fringe may be only 10 to 15 cm. In a loamy or clayey soil that does not shrink or swell appreciably, the thickness may be 30 cm or more, depending on the size distribution of the pores.

The duration of the period that the soil must be saturated to have an aquic regime is not known. The duration must be at least a few days, because it is implicit in the concept that dissolved oxygen is virtually absent. Because dissolved oxygen is removed from ground water by respiration of micro-organisms, roots and soil fauna, it is also implicit in the concept that the soil temperature is above biologic zero (5°C) at some time while the soil or the horizon is saturated.

Very commonly, the level of ground water fluctuates with the seasons. The level is highest in the rainy season, or in fall, winter, or spring if cold weather virtually stops evapotranspiration. There are soils, however, in which the ground water is always at or very close to the surface. A tidal marsh and a closed, landlocked depression fed by perennial streams are examples. The moisture regime in these soils is called "peraquic." Although the term is not used as a formative element for names of taxa, it is used in their descriptions as an aid in understanding genesis.

Aridic and torric (L. *aridus*, dry, and L. *torridus*[6], hot and dry) **moisture regimes.**--These terms are used for the same moisture regime but in different categories of the taxonomy.

In the aridic (torric) moisture regime, the moisture control section in most years is

1. Dry in all parts more than half the time (cumulative) that the soil temperature at a depth of 50 cm is above 5°C; and

2. Never moist in some or all parts for as long as 90 consecutive days when the soil temperature at a depth of 50 cm is above 8°C.

Soils that have an aridic or a torric moisture regime are normally in arid climates. A few are in semiarid climates and either have physical properties that keep them dry, such as a crusty surface that virtually precludes infiltration of water, or they are very shallow over bedrock. There is little or no leaching in these moisture regimes, and soluble salts accumulate in the soil if there is a source of them.

The limits of soil temperature exclude from these moisture regimes the very cold and dry regions of Greenland and adjacent islands. Such fragmentary data are available on the soils of those regions that no provision is made for their moisture regimes in this taxonomy.

Udic moisture regime.--The udic (L. *udus*, humid) moisture regime implies that in most years the soil moisture control section is not dry in any part for as long as 90 days (cumulative). If the mean annual soil temperature is lower than 22°C and if the mean winter and mean summer soil temperatures at a depth of 50 cm differ by 5°C or more, the soil moisture control section is not dry in all parts for as long as 45 consecutive days in the 4 months that follow the summer solstice in 6 or more years out of 10. In addition, the udic moisture regime requires, except for short periods, a three-phase system, solid-liquid-gas, in part, but not necessarily in all, of the soil when the soil temperature is above 5°C.

The udic moisture regime is common to the soils of humid climates that have well-distributed rainfall or that have enough rain in summer that the amount of stored moisture plus rainfall is approximately equal to or exceeds the amount of evapotranspiration. Water moves down through the soil at some time in most years.

---

[6]Torridus is not an ideal root, but a better one could not be found. Although soils may not be hot throughout the year, soils that have a torric moisture regime are hot and dry in summer.

If precipitation exceeds evapotranspiration in all months of
most years, there are occasional brief periods when some
stored moisture is used, but the moisture tension rarely becomes as great as 100 kPa in the soil moisture control section. The water moves through the soil in all months that it
is not frozen. This extremely wet moisture regime is called
"perudic" (L. *per*, throughout in time, L. *udus*, moist). The
formative element *ud* is used in the names of most taxa to
indicate either a udic or a perudic regime. The formative
element *per* is used in selected taxa.

Ustic moisture regime.--The ustic (L. *ustus*, burnt, implying
dryness) moisture regime is intermediate between the aridic
and the udic regime. The concept is one of limited moisture, but the moisture is present at a time when conditions
are suitable for plant growth. The ustic moisture regime is
not applied to soils that have cryic or pergelic temperature
regimes, which are defined later.

If the mean annual soil temperature is 22°C or higher or if
the mean summer and winter soil temperatures differ by
<5°C at a depth of 50 cm, the soil moisture control section
in the ustic moisture regime is dry in some or all parts for
90 or more cumulative days in most years. But the moisture
control section is moist in some part for more than 180 cumulative days, or it is continuously moist in some part for at
least 90 consecutive days.

If the mean annual soil temperature is lower than 22°C and
if the mean summer and winter soil temperatures differ by
5°C or more at a depth of 50 cm, the soil moisture control
section in the ustic regime is dry in some or all parts for 90
or more cumulative days in most years. But it is not dry in
all parts for more than half the time that the soil temperature is higher than 5°C at a depth of 50 cm (the aridic and
torric regimes). Also, it is not dry in all parts for as long as
45 consecutive days in the 4 months that follow the summer
solstice in 6 or more years out of 10 if the moisture control
section is moist in all parts for 45 or more consecutive days
in the 4 months that follow the winter solstice in 6 or more
years out of 10 (xeric regime).

In tropical and subtropical regions that have either one or
two dry seasons, summer and winter have little meaning. In
those regions, the ustic regime is that typified in a monsoon
climate that has at least one rainy season of 3 months or
more. In temperate regions of subhumid or semiarid climates, the rainy seasons are usually spring and summer or
spring and fall, but never winter. Native plants are mostly
annuals or they have a dormant period while the soil is dry.

Xeric moisture regime.--The xeric moisture regime (Gr.
xeros, dry) is that typified in Mediterranean climates, where
winters are moist and cool and summers are warm and dry.
The moisture, coming in winter when potential evapotranspiration is at a minimum, is particularly effective for
leaching. In a xeric moisture regime, the soil moisture control section is dry in all parts for 45 or more consecutive
days within the 4 months that follow the summer solstice in
6 or more years out of 10. It is moist in all parts for 45 or
more consecutive days within the 4 months that follow the

winter solstice in 6 or more years out of 10. The moisture control section is moist in some part more than half the time, cumulative, that the soil temperature at a depth of 50 cm is higher than 5°C, or in 6 or more years out of 10 it is moist in some part for at least 90 consecutive days when the soil temperature at a depth of 50 cm is continuously higher than 8°C. In addition, the mean annual soil temperature is lower than 22°C, and mean summer and mean winter soil temperatures differ by 5°C or more at a depth of 50 cm or at a lithic or paralithic contact, whichever is shallower.

Soil temperature regimes

<u>Classes of soil temperature regimes</u>

The following soil temperature regimes are used in defining classes at various categoric levels in the taxonomy.

Pergelic (L. *per*, throughout in time and space, and L. *gelare*, to freeze; connoting permanent frost)..--Soils with a pergelic temperature regime have a mean annual temperature lower than 0°C. These are soils that have permafrost if they are moist, or dry frost if excess water is not present. It seems likely that the moist and the dry pergelic regimes should be defined separately, but at present we have only fragmentary data on the dry soils of very high latitudes.

Cryic (Gr. *kryos*, coldness; connoting very cold soils).--In this regime soils have a mean annual temperature higher than 0°C (32°F) but lower than 8°C (47°F).

1. In mineral soils the mean summer soil temperature (June, July, and August in the northern hemisphere and December, January, and February in the southern hemisphere) at a depth of 50 cm or at a lithic or paralithic contact, whichever is shallower, is as follows:
    a. If the soil is not saturated with water during some part of the summer and
        (1) There is no O horizon, lower than 15°C (59°F);
        (2) There is an O horizon, lower than 8°C (47°F);
    b. If the soil is saturated with water during some part of the summer and
        (1) There is no O horizon, lower than 13°C (55°F);
        (2) There is an O horizon or a histic epipedon, lower than 6°C (43°F).

2. In organic soils, either
    a. The soil is frozen in some layer within the control section in most years about 2 months after the summer solstice; that is, the soil is very cold in winter but warms up slightly in summer; or
    b. The soil is not frozen in most years below a depth of 5 cm; that is, the soil is cold throughout the year but, because of marine influence, does not freeze in most years.

Cryic soils that have an aquic moisture regime commonly are churned by frost.

Most isofrigid soils with a mean annual soil temperature above 0°C have a cryic temperature regime. A few with organic materials in the upper part are exceptions. Throughout this text all isofrigid soils without permafrost are considered to have a cryic temperature regime.

Frigid.--The frigid regime and some of the others that follow are used chiefly in defining classes of soils in the low categories. In the frigid regime the soil is warmer in summer than one in the cryic regime, but its mean annual temperature is lower than 8°C (47°F), and the difference between mean winter and mean summer soil temperature is more than 5°C (9°F) at a depth of 50 cm or at a lithic or paralithic contact, whichever is shallower.

Mesic.--The mean annual soil temperature is 8°C or higher but lower than 15°C (59°F) and the difference between mean summer and mean winter soil temperature is more than 5°C at a depth of 50 cm or at a lithic or paralithic contact, whichever is shallower.

Thermic.--The mean annual soil temperature is 15°C (59°F) or higher but lower than 22°C (72°F), and the difference between mean summer and mean winter soil temperature is more than 5°C at a depth of 50 cm or at a lithic or paralithic contact, whichever is shallower.

Hyperthermic.--The mean annual soil temperature is 22°C (72°F) or higher, and the difference between mean summer and mean winter soil temperature is more than 5°C at a depth of 50 cm or at a lithic or paralithic contact, whichever is shallower.

If the name of a soil temperature regime has the prefix iso, the mean summer and winter soil temperature for June, July, and August and for December, January, and February differ by less than 5°C at a depth of 50 cm or at a lithic or paralithic contact, whichever is shallower.

Isofrigid.--The mean annual soil temperature is lower than 8°C (47°F).

Isomesic.--The mean annual soil temperature is 8°C or higher but lower than 15°C (59°F).

Isothermic.--The mean annual soil temperature is 15°C or higher but lower than 22°C (72°F).

Isohyperthermic.--The mean annual soil temperature is 22°C or higher.

The limit between isofrigid and isomesic cannot be tested in the United States and is tentative.

Sulfidic materials

Sulfidic materials are waterlogged mineral or organic soil materials that contain 0.75 percent or more sulfur (dry weight), mostly in the form of sulfides and that have less than three times as much carbonate ($CaCO_3$ equivalent) as sulfur. Sulfidic materials accumulate in a soil that is permanently saturated, generally with brackish water. The sulfates in the water are biologically reduced to sulfides as the soil materials accumulate. Sulfidic materials are most common

in coastal marshes near the mouths of rivers that carry
noncalcareous sediments, but they may occur in fresh-water
marshes if there is sulfur in the water. If the soil is
drained, the sulfides oxidize and form sulfuric acid. The
pH, which normally is near neutrality before drainage, may
drop below 2. The acid reacts with the soil to form iron
and aluminum sulfates. The iron sulfate, jarosite, segregates
and forms the bright-yellow mottles that characterize a sulfuric horizon. The transition from sulfidic materials to a
sulfuric horizon normally requires a very few years. A
sample of sulfidic materials, if air dried slowly in shade for
about 2 months with occasional remoistening, becomes extremely acid. For quick identification in the field, a sample
can be oxidized by boiling in concentrated H_2O_2 and measuring the drop in pH[7].

Thixotropy

Thixotropy is "a reversible gel-sol transformation under
isothermal shearing stress following rest" (Webster's 1967).
The term means "to change by touch". Many kinds of
thixotropic substances have been identified and studied, including some sesquioxide gels, kaolinite gels, montmoril-
lonite gels, greases, inks, paints, protoplasm, blood coagula,
nitrocellulose solutions, and drilling muds. Thixotropy ap-
parently is the result of a kind of structure that, if broken
down, can rebuild itself. The breakdown may be caused by
one of several actions: by agitation, by shearing, or even by
ultrasonic waves. Some natural (untreated) soil materials ex-
hibit this property. A field test of thixotropic soil is this:
Press a bit of wet soil between thumb and forefinger; at
first it resists deformation, having some rigidity, or elastic-
ity, or both; under increasing pressure the soil can be
molded and deformed; under greater pressure, suddenly the
soil changes from a plastic solid to a liquid, and the fingers
skid. After the soil smears in this fashion, usually free water can be seen on the fingers. In a matter of a second or
two the liquefied soil sets again to its original solid state. If
a knife blade is pushed into the soil mass in a pit and removed suddenly, it has only a staining of muddy water; if
pressed into the soil and slowly pulled out, a large mass of
soil adheres to the blade. In the literature of soils of western United States, particularly of Hawaii, the consistence
term "smeary" is used to characterize soil materials that are
thixotropic.

Tonguing and interfingering

<u>Tonguing of albic materials</u>

Tongues of albic materials consist of penetrations of
bleached material that has the color of an albic horizon in
an argillic or a natric horizon, along ped surfaces if peds are
present. No continuous albic horizon need be present above
the tongues. The penetrations have a vertical dimension of
>5 cm in any argillic or natric horizon. Their horizontal
dimension is 5 mm or more in a fine-textured argillic or

---

[7]Concentrated H_2O_2 can cause serious burns and is dangerous. Gloves should
be worn, and precautions should be taken against spilling, leakage, or spat-
tering.

natric horizon (clay, silty clay, or sandy clay), 10 mm or more in a moderately fine textured argillic or natric horizon (clay loam, sandy clay loam, or silty clay loam), and 15 mm or more in a medium or coarser textured argillic or natric horizon (silt loam, loam, very fine sandy loam, or coarser). The penetrations must occupy more than 15 percent of the matrix of some part of the argillic or natric horizon before they are considered tongues.

<u>Interfingering of albic materials</u>

Interfingering of albic materials consists of penetrations of albic materials into an underlying argillic or natric horizon along faces of peds, primarily along vertical faces but to a lesser degree along horizontal faces. No continuous albic horizon need be present. The penetrations are not wide enough to constitute tonguing, but they form continuous skeletans (ped coatings of clean silt or sand defined by Brewer, 1964) >1 mm thick on the vertical ped faces, which means a total width >2 mm between abutting peds. Because quartz is such a common constituent of soils, the skeletans usually appear to be nearly white when dry and light gray when moist, but their color is determined in large part by the color of the sand or silt fraction.

To be recognized as interfingering, all the following requirements must be met in a horizon that is 5 cm or more thick:

1. Half or more of the matrix consists of peds of the argillic or natric horizon;

2. Albic materials are thicker than 2 mm on vertical faces between abutting peds but are too thin to be tongues;

3. Clay skins are present in the peds, at least in pores.

Albic materials meet the following requirements for color. If the value, dry, is 7 or more, or the value, moist, is 6 or more, the chroma is 3 or less either dry or moist. If the value, dry, is 5 or 6 and the value, moist, is 4 or 5, the chroma is closer to 2 than to 3 either dry or moist.

Weatherable minerals

Several references are made to weatherable minerals in the text of this chapter and later chapters. Obviously, the stability of a mineral in a soil is a partial function of the soil moisture regime. In the context of the references in the definitions of diagnostic horizons and of various taxa, a humid climate is always assumed, either present or past. Minerals that are included in the meaning of weatherable minerals are:

1. Clay minerals: All 2:1 lattice clays except one that is currently considered to be an aluminum-interlayered chlorite. Sepiolite, talc, and glauconite are also included in the meaning of this group of weatherable clay minerals, although they are not everywhere of clay size.

2. Silt- and sand-size minerals (0.02 to 0.2 mm in diameter): Feldspars, feldspathoids, ferromagnesian minerals, glass, micas, zeolites, and apatite.

It should be noted that this is a restricted meaning of weatherable minerals. Calcite, for example, is readily soluble in a humid environment. If it is dissolved, it leaves no trace or residue. Soils that have been intensely and deeply weathered in a humid environment of the past are, in some places, preserved today in an arid environment. Calcite could reappear in one of these soils if it were brought in as dust. The intent is to include, in the context of the meaning of weatherable minerals for this purpose, only those minerals that are unstable in a humid climate relative to other minerals, such as quartz and 1:1 lattice clays, and that are more resistant to weathering than calcite.

LITERATURE CITED

Brewer, R. 1964. Fabric and mineral analysis of soils. John Wiley and Sons, Inc., New York, London, Sydney. 470 p.

Pons, L.J. and I.S. Zonneveld. 1965. Soil ripening and soil classification. Initial soil formation in alluvial deposits and a classification of the resulting soils. Int. Inst. Land Reclam. and Impr. Pub. 13. Wageningen, The Netherlands. 128 p.

Chapter 2

Horizons and Properties Diagnostic for the Higher Categories; Organic Soils

Organic Soil Materials

Organic soil materials and organic soils:

1. Are saturated with water for long periods or are artificially drained and, excluding live roots, (a) have 18 percent or more organic carbon if the mineral fraction is 60 percent or more clay, (b) have 12 percent or more organic carbon if the mineral fraction has no clay, or (c) have a proportional content of organic carbon between 12 and 18 percent if the clay content of the mineral fraction is between zero and 60 percent; or

2. Are never saturated with water for more than a few days and have 20 percent or more organic carbon.

Definition Of Organic Soils

Organic soils (Histosols) are soils that

1. Have organic soil materials that extend from the surface to one of the following:
   a. A depth within 10 cm or less of a lithic or paralithic contact, provided the thickness of the organic soil materials is more than twice that of the mineral soil above the contact; or
   b. Any depth if the organic soil material rests on fragmental material (gravel, stones, cobbles) and the interstices are filled with organic materials, or rests on a lithic or paralithic contact; or

2. Have organic materials that have an upper boundary within 40 cm of the surface and
   a. Have one of the following thicknesses:
      (1) 60 cm or more if three-fourths or more of the volume is moss fibers or the moist bulk density is <0.1 g per cubic centimeter (6.25 lbs per cubic foot);
      (2) 40 cm or more if
         (a) The organic soil material is saturated with water for long periods (>6 months) or is artificially drained; and
         (b) The organic material consists of sapric or hemic materials or consists of fibric materials that are less than three-fourths moss fibers by volume and a moist bulk density of 0.1 or more; and
   b. Have organic soil materials that
      (1) Do not have a mineral layer as much as 40 cm thick either at the surface or whose upper boundary is within a depth of 40 cm from the surface; and
      (2) Do not have mineral layers, taken cumulatively, as thick as 40 cm within the upper 80 cm.

It is a general rule that a soil is classed as an organic soil (Histosol) either if more than half of the upper 80 cm (32 in.) of soil is organic or if organic soil material of any thickness rests on rock or on fragmental material having interstices filled with organic materials.

Kinds Of Organic Soil Materials

Three basic kinds of organic soil materials are distinguished, fibric, hemic, and sapric, according to the degree of decomposition of the original plant materials.

Fibers

A fiber is a fragment or piece of plant tissue, excluding live roots, that is large enough to be retained on a 100-mesh sieve (openings 0.15 mm in diameter) and that retains recognizable cellular structure of the plant from which it came. The material is screened after dispersion in sodium hexametaphosphate. Fragments larger than 2 cm in cross section or in their smallest dimension, to be called fibers, must be decomposed enough that they can be crushed and shredded with the fingers. Fragments of wood that are larger than 2 cm in cross section and that are so undecomposed that they cannot be crushed and shredded with the fingers are not considered fibers.

Fibric soil materials (L. *fibra*, fiber)

Fibric soil materials have the following characteristics:

1. The fiber content after rubbing is three-fourths[1] or more of the soil volume, excluding coarse fragments and mineral layers; or

2. The fiber content after rubbing is two-fifths or more of the soil volume, excluding coarse fragments and mineral layers, and the material yields a sodium pyrophosphate extract color on white chromatographic paper that has value and chroma of 7/1, 7/2, 8/1, 8/2, or 8/3.

Hemic soil materials

Hemic soil materials (Gr. *hemi*, half; implying intermediate decomposition) are intermediate in degree of decomposition between the less decomposed fibric and the more decomposed sapric materials. They have morphological features that give intermediate values for fiber content, bulk density, and water content. They are partly altered both physically and biochemically.

Sapric soil materials (Gr. *sapros*, rotten)

These are the most highly decomposed of the organic materials. They normally have the smallest amount of plant fiber, the highest bulk density, and the lowest water content on a dry-weight basis at saturation. They are commonly very dark gray to black. They are relatively stable, i.e., they change very little physically and chemically with time in comparison to the others.

---

[1]Fractions are used rather than percentages to avoid implying a higher degree of accuracy than is warranted.

Sapric materials have the following characteristics:

1. The fiber content after rubbing is less than one-sixth of the soil volume, excluding coarse fragments and mineral layers; and

2. The sodium pyrophosphate extract color on chromatographic paper is below or to the right of a line drawn to exclude blocks 5/1, 6/2, and 7/3 on the chart.

Humilluvic materials

Illuvial humus accumulates in the lower parts of some organic soils if they are acid and have been drained and cultivated. The illuvial humus has a younger C^{14} age than the overlying organic materials. It has very high solubility in sodium pyrophosphate and rewets very slowly after drying. Most commonly it accumulates near a contact with a sandy mineral horizon. To be recognized as a differentia in classification, the illuvial humus should constitute at least half the volume of a layer at least 2 cm thick.

Limnic materials

Limnic materials include both organic and inorganic materials that were either (1) deposited in water by precipitation or through the action of aquatic organisms such as algae or diatoms, or (2) derived from underwater and floating aquatic plants and subsequently modified by aquatic animals. They include coprogenous earth (sedimentary peat), diatomaceous earth, and marl.

Coprogenous earth

A coprogenous earth (sedimentary peat) layer is a limnic layer that

1. Contains many fecal pellets a few hundredths to a few tenths of a millimeter in diameter;

2. Has a color value, moist, <5;

3. Either forms a slightly viscous water suspension and is slightly plastic but not sticky or shrinks upon drying to form clods that are difficult to rewet and that often tend to crack along horizontal planes;

4. Is normally but not necessarily nearly devoid of fragments of plants that can be recognized with the eye; and

5. Yields a saturated sodium pyrophosphate extract on white filter paper that has higher color value and lower chroma than 10 YR 7/3 or the cation-exchange capacity is <240 cmol(+) kg^{-1} of organic matter (measured by loss on ignition) or both.

Diatomaceous earth

A diatomaceous earth layer is a limnic layer that

1. Has a matrix color value of 3 through 5 if not previously dried, and the value changes irreversibly on drying. The color change results from irreversible shrinkage of organic matter coatings on diatoms, which can be identified by microscopic (440X) examination of dry samples; and

2. Yields a color higher in value and lower in chroma than 10 YR 7/3 on white filter paper that is inserted into a paste made of the material in a saturated sodium pyrophosphate solution or the cation-exchange capacity is <240 cmol(+) kg^{-1} of organic matter (by loss on ignition) or both.

Marl

A marl layer is a limnic layer that

1. Has a color value, moist, of 5 or more; and

2. Reacts with dilute HCl to evolve CO_2. Marl usually does not change color irreversibly on drying. A layer of marl contains too little organic matter to coat the carbonate, even before it has been shrunk by drying.

Thickness Of Organic Materials (control section)

For practical reasons an arbitrary control section has been established for taxonomy of Histosols. It is either 130 cm (51 in.) or 160 cm (63 in.) thick, depending on the kind of material, provided that no lithic or paralithic contact, thick layer of water, or frozen soil occurs within those limits. The thicker control section is used if the surface layer to a depth of 60 cm (24 in.) has three-fourths or more fibers derived from *Sphagnum* or from *Hypnum* or other mosses or has a bulk density <0.1. Layers of water may be thin or thick, from a few centimeters to many meters. Water is taken as the base of the control section only if the water extends below a depth of 130 cm or 160 cm, depending on the kind of material above it. A lithic or a paralithic contact shallower than 130 cm (51 in.) or 160 cm (63 in.), depending on the kind of material above it, is taken as the base of the control section, or the base is placed 25 cm (10 in.) below the depth at which the soil is frozen about 2 months after the summer solstice. An unconsolidated mineral substratum shallower than those limits does not change the base of the control section.

The control section has been divided somewhat arbitrarily into three tiers, the surface, subsurface, and bottom tiers.

Surface tier

The surface tier is the upper 60 cm (24 in.) if (1) the materials is fibric and three-fourths or more of the fiber volume is derived from *Sphagnum* or mosses or (2) the material has a bulk density <0.1 g cm^{-3}; otherwise, the surface tier is the top 30 cm (12 in.) exclusive of loose surface litter or living mosses.

A surface mineral layer <40 cm (16 in.) thick is present on some organic soils as a result of flooding, additions by men to increase soil strength or reduce frost hazard, volcanic eruptions, or other causes. If present, it is considered a part of the surface tier, even though it may be >30 cm thick, and the depth then is measured from the top of the mineral layer.

Subsurface tier

The subsurface tier is 60 cm (24 in.) thick unless the control section ends at a lithic or paralithic contact or at water within this depth or unless the soil is frozen at too shallow a depth. In any of these situations the subsurface tier extends from the base of the surface tier to the base of the control section. It includes any unconsolidated mineral layers that may be present within those depths.

Bottom tier

The bottom tier is 40 cm (16 in.) thick unless the control section stops within the maximum span.

Chapter 3

Family Differentiae

Mineral Soils

The differentiae used to distinguish families of mineral soils within a subgroup are listed next in the order in which the descriptive terms appear in the family name and in which the terms are defined in this chapter.

>Particle-size classes
>Mineralogy classes
>Calcareous and reaction classes
>Soil temperature classes
>Soil depth classes
>Soil slopes classes
>Soil consistence classes
>Classes of coatings (on sand)
>Classes of cracks

Particle-size classes

Particle size refers to grain-size distribution of the whole soil and is not the same as texture, which refers to the fine-earth fraction. The fine-earth fraction consists of the particles that have a diameter <2 mm. Particle-size classes are a kind of compromise between engineering and pedologic classifications. The limit between sand and silt is a diameter of 74 micrometers in the engineering classification and of either 50 or 20 micrometers in pedologic classifications. The engineering classifications are based on percentages by weight in the fraction <74 mm in diameter, and textural classes are based on percentages by weight in the fraction <2 mm in diameter.

The very fine sand separate (diameter between 0.05 mm and 0.1 mm) is split in engineering classifications. In defining particle-size classes, much the same split is made but in a different manner. A fine sand or loamy fine sand normally has an appreciable content of very fine sand, but the very fine sand fraction is mostly coarser than 74 micrometers. A silty sediment, such as loess, may also have an appreciable component of very fine sand, but most of the very fine sand is finer than 74 micrometers. So, in particle-size classes, the very fine sand is allowed to "float." It is treated as sand if the texture is fine sand, loamy fine sand, or a coarser class. It is treated as silt if the texture is very fine sand, loamy very fine sand, sandy loam, silt loam, or a finer class.

No single set of particle-size classes seems appropriate as family differentiae for all kinds of soils. The classes that follow provide for a choice of either 7 or 11 particle-size classes. This choice permits relatively fine distinctions in soils if the particle size is important and broader groupings if the particle size is not susceptible to precise measurement or if the use of narrowly defined classes produces undesirable groupings. Thus in some families the term "clayey" indicates that there is 35 percent or more clay in defined horizons, but in other families the term "fine" indicates that the clay fraction constitutes 35 through 59 percent of the fine earth of the horizons, and the term "very-fine" indicates 60 percent or more clay. The term "rock fragments" refers to particles 2 mm in diameter or larger and includes all sizes that have horizontal dimensions less than the size of a pedon. It is not the same as coarse fragments, which excludes stones and boulders larger than about 25 cm. The term "fine earth" refers to particles smaller than 2 mm in diameter.

Definition of classes

Fragmental.--Stones, cobbles, gravel, and very coarse sand particles; too little fine earth to fill some of the interstices larger than 1 mm in diameter.

Sandy-skeletal.--Rock fragments 2 mm in diameter or larger make up 35 percent or more by volume; enough earth to fill interstices larger than 1 mm; the fraction finer than 2 mm is sandy as defined for the sandy particle-size class.

Loamy-skeletal.--Rock fragments make up 35 percent or more by volume; enough fine earth to fill interstices larger than 1 mm; the fraction finer than 2 mm is loamy as defined for the loamy particle-size class.

Clayey-skeletal.--Rock fragments make up 35 percent or more by volume; enough fine earth to fill interstices larger than 1 mm; the fraction finer than 2 mm is clayey as defined for the clayey particle-size class.

Sandy.--The texture of the fine earth is sand or loamy sand that contains less than 50 percent very fine sand; rock fragments make up less than 35 percent by volume.

Loamy[1].--The texture of the fine earth is loamy very fine sand, very fine sand, or finer, but the amount of clay is <35 percent; rock fragments are <35 percent by volume.
Coarse-loamy. By weight, 15 percent or more of the particles are fine sand (diameter 0.25-0.1 mm) or coarser, including fragments up to 7.5 cm in diameter; <18 percent clay in the fine-earth fraction.
Fine-loamy. By weight, 15 percent or more of the particles are fine sand (diameter 0.25-0.1 mm) or coarser, including fragments up to 7.5 cm in diameter; 18 through 34 percent clay in the fine-earth fraction (<30 percent in Vertisols).
Coarse-silty. By weight, <15 percent of the particles are fine sand (diameter 0.25-0.1 mm) or coarser, including fragments up to 7.5 cm in diameter; <18 percent clay in the fine-earth fraction.
Fine-silty. By weight, <15 percent of the particles are fine sand (diameter 0.25-0.1 mm) or coarser, including fragments up to 7.5 cm in diameter; 18 through 34 percent clay in the fine-earth fraction (<30 percent in Vertisols).

---

[1]If the ratio of 1500 kPa water to clay is 0.6 or more in half or more of the control section, the percentage of clay is considered to be 2.5 times the percentage of 1500 kPa water. Carbonates of clay size are not considered to be clay but are treated as silt in all particle-size classes.

<u>Clayey</u>[2].-- The fine earth contains 35 percent or more clay
by weight, and rock fragments are <35 percent by volume.
Fine. A clayey particle-size class that has 35 through 59
percent clay in the fine-earth fraction (30 through 50 percent in Vertisols).
Very-fine. A clayey particle-size class that has 60 percent
or more clay in the fine-earth fraction.

<u>Modifiers that replace names of particle-size classes</u>

There are three situations in which particle-size class names
are not used. In one, the name is redundant. Psamments
and Psammaquents, by definition, are sandy, and no particle-size class name is needed or used in the family name.

In the second situation, particle size is a meaningless concept
because, presumably, the soil consists of a mixture of discrete mineral particles and of gels. The concept of either
texture or particle size is not applicable to a gel, particularly
if the gel cannot be dispersed. Consequently, no particle-
size class names are used if the soil is mostly glass or if the
exchange complex is dominated by amorphous materials, as
is the situation with Andepts by definition. In families of
Andepts and Andaquepts, in most andic subgroups of Inceptisols and andaqueptic and andeptic subgroups of other
orders, and in families of Entisols and Aridisols with modifiers listed below, particle-size class names, as such, are not
used for the part of the soil that does not disperse.

In the third situation, the content of organic matter is high
and particle size has only limited relation to the physical and
chemical properties of the soils. This seems to be normal in
soils that have both a cryic temperature regime and a spodic
horizon. Therefore, particle-size class names are not used
for the spodic horizons[3] or Cryaquods, Cryohumods, Cryorthods, or Cryic Placohumods.

The following terms are substituted for the particle-size
class names in the taxa that have been listed unless the particle-size modifier is redundant. They reflect a combination
of particle size and mineralogy, and they take the place of
both.

<u>Cindery.</u>--Sixty percent or more of the whole soil (by
weight)[4] volcanic ash, cinders, and pumice; 35 percent or
more (by volume) is cinders that have diameter of 2 mm or
more.

---

[2]If the ratio of 1500 kPa water to clay is 0.6 or more in half or more of the
control section, the percentage of clay is considered to be 2.5 times the percentage of 1500 kPa water. Carbonates of clay size are not considered to be
clay but are treated as silt in all particle-size classes.

[3]Particle-size class names are applied to other spodic horizons but with reser-
vations. Somewhat different classes probably should be used for most families
of Spodosols, but data are too few to permit the testing of alternatives. Some
series that would otherwise be reasonably homogeneous are split at the family
level by the particle-size classes. These soils have appreciable but not very
large amounts of organic matter in the spodic horizon.

[4]Percentages by weight in these definitions are estimated from grain counts;
usually, a count of one or two dominant size fractions of the conventional
mechanical analysis is enough for placement of the soil.

<u>Ashy and ashy-skeletal.</u>--These are mainly soils that have a fine-earth fraction that feels like a sand or a loamy sand after prolonged rubbing.

Ashy. Sixty percent or more of the whole soil (by weight) volcanic ash, cinders, and pumice; <35 percent (by volume) is 2 mm in diameter or larger.

Ashy-skeletal. Rock fragments other than cinders are 35 percent or more (by volume); the fine-earth fraction is otherwise ashy.

<u>Medial and medial-skeletal.</u>--These are soils that have a fine-earth fraction that feels loamy, as the term is defined earlier in this chapter, after prolonged rubbing.

Medial. Less than 60 percent of the whole soil (by weight) volcanic ash, cinders, and pumice; <35 percent (by volume) is 2 mm if diameter or larger; the fine-earth fraction is not thixotropic; the exchange complex is dominated by amorphous materials.

Medial-skeletal. Thirty-five percent (by volume) or more rock fragments other than cinders 2 mm in diameter or larger; the fine-earth fraction is otherwise medial.

<u>Thixotropic and thixotropic-skeletal.</u>--These are soils that have a fine-earth fraction that is thixotropic and an exchange complex dominated by amorphous clays.

Thixotropic. Less than 35 percent (by volume) has diameter of 2 mm or larger; the fine-earth fraction is thixotropic; the exchange complex is dominated by amorphous materials.

Thixotropic-skeletal. Thirty-five percent or more (by volume) rock fragments other than cinders 2 mm in diameter or larger; the fine-earth fraction is thixotropic.

<u>Control section for particle-size classes or their substitutes</u>

Names of particle-size classes or their substitutes as defined are not applied to a fragipan, duripan, or petrocalcic horizon but are applied to specific horizons or to the materials between given limits of depth that are defined in terms of either the distance below the surface of the mineral soil or the upper boundary of a specified horizon or root-limiting layer. The vertical section so defined is called the control section. Definitions of the control section for determination of the particle-size classes are arranged as a key.

A. Particle-size modifiers or substitutes are used to describe material from the surface to a lithic or paralithic contact, or to a fragipan, duripan, or petrocalcic horizon if any of these come within a depth of 36 cm (14 in.) or less; or to a depth of 36 cm if the soil temperature is 0°C or lower within this depth about 2 months after the summer solstice.

B. In other soils that do not have an argillic or kandic horizon or a natric horizon and in great groups of Spodosols, Alfisols, and Ultisols that have a spodic horizon or a fragipan in or above an argillic horizon:

   1. Particle-size modifiers or substitutes are used to describe material from the base of the Ap horizon or from a depth of 25 cm, whichever is greater, to a root-limiting layer if the depth is <1 m; or to a depth 25 cm below the level at which the soil temperature is 0 degree

C about 2 months after the summer solstice; whichever is
shallower.
2. Otherwise, particle-size modifiers or substitutes are
used to describe material from a depth of 25 cm to a
depth of 1 m.

C. In other soils of the orders Alfisols and Ultisols and in
great groups of Aridisols and Mollisols that have an argillic
or kandic horizon that has (a) a lower boundary deeper than
25 cm (see E) and (b) an upper boundary shallower than 1
m, or the soil is in a grossarenic subgroup:
1. If there are no strongly contrasting particle-size
classes, as defined later, and there is no root-limiting
layer between the top of the argillic, kandic, or natric
horizon and a depth of 1 m, particle-size modifiers or
substitutes are used to describe the whole argillic or
natric horizon if it is <50 cm thick[5] or the upper 50 cm
of the argillic, kandic, or natric horizon if it is >50 cm
thick.
2. If there are horizons or layers of strongly contrasting
particle-size classes, as defined later, within or below the
argillic, kandic, or natric horizon and within a depth of
1 m, particle-size modifiers or substitutes are used to
describe material from the top of the argillic, kandic, or
natric horizon to a depth of 1 m or to a lithic or
paralithic contact, duripan, fragipan, or petrocalcic
horizon, whichever is shallower.
3. If there is a root-limiting layer below an argillic,
kandic, or natric horizon, particle-size modifiers or
substitutes are used to describe material from the top of
the argillic horizon, excluding any part incorporated in an
Ap horizon, to the top of the fragipan, duripan, or
petrocalcic horizon, or are used to describe the upper 50
cm of the argillic, kandic, or natric horizon whichever
of these is less.

D. In other soils in the orders Alfisols and Ultisols and in
great groups of Aridisols and Mollisols that have an argillic,
kandic, or natric horizon that has its upper boundary at a
depth >1 m and that are not in a grossarenic subgroup, particle-size modifiers or substitutes are applied to describe
material from a depth of 25 cm to a depth of 1 m below the
mineral surface.

E. In other soils in which the lower boundary of the argillic
or natric horizon is shallower than 25 cm, that is, they have
a *k* horizon in which there is soft powdery lime, or have a
calcic or other named diagnostic horizon that has its upper
boundary within 25 cm of the surface, or have rock structure dominant within that depth, particle-size classes are
used to describe material from the top of the argillic horizon
or the base of an Ap horizon, whichever is shallower, to a

------

[5]The upper boundary of the argillic or kandic horizon is not always obvious.
If properties of an argillic horizon are present but the upper boundary is
gradual, use the depth at which the percentage of clay exceeds that of a
higher lying horizon by the appropriate amount after fitting to a smooth
curve. If the boundary is irregular or broken, as in A&B or B&A horizons, use
the depth at which half or more of the volume has the fabric of an argillic
horizon.

lithic or paralithic contact, a petrocalcic or petrogypsic horizon, duripan, or to a depth of 1 m, whichever is shallowest.

<u>Strongly contrasting particle-size classes</u>

In applying names of particle-size classes, the weighted average particle-size class of the control section or of the horizon listed is named unless there are strongly contrasting particle-size classes within the control section or the horizons. If there are strongly contrasting particle-size classes, both particle-size classes are named. Thus, if the weighted average of the upper part of the control section is loamy fine sand and the lower part is clay, the family differentia is sandy over clayey. If there are more than two contrasting particle-size classes within the control section, the classes differing most in median particle size are named. Sandy includes fine sand as well as coarser sands. Medial, ashy, or thixotropic substitutes are applied only if the materials extend at least 10 cm into the upper part of the control section.

The following particle-size classes are strongly contrasting if the transition between them is less than 12.5 cm thick:
1. Cindery over sandy or sandy-skeletal.
2. Cindery over loamy.
3. Sandy-skeletal over loamy if the loamy material has <50 percent fine or coarser sand.
4. Sandy over loamy if the loamy material has <50 percent fine or coarser sand.
5. Sandy over clayey.
6. Ashy over cindery.
7. Ashy over loamy-skeletal.
8. Ashy over loamy.
9. Loamy-skeletal over fragmental.
10. Loamy-skeletal over sandy.
11. Loamy-skeletal over clayey if there is an absolute difference of >25 percent in the percentages of clay in the fine-earth fractions.
12. Clayey-skeletal over sandy.
13. Medial over fragmental.
14. Medial over cindery.
15. Medial over sandy or sandy-skeletal.
16. Medial over loamy-skeletal.
17. Medial over loamy.
18. Medial over clayey.
19. Medial over thixotropic.
20. Coarse-loamy over fragmental.
21. Coarse-loamy over sandy or sandy-skeletal if the coarse-loamy material has <50 percent fine or coarser sand.
22. Loamy over sandy or sandy-skeletal if the loamy material has <50 percent fine or coarser sand.
23. Coarse-loamy over clayey.
24. Coarse-silty over sandy or sandy-skeletal.
25. Coarse-silty over clayey.
26. Fine-loamy over fragmental.
27. Fine-loamy over sandy or sandy-skeletal.
28. Fine-loamy over clayey if there is an absolute difference of >25 percent in the percentage of clay.
29. Fine-loamy over cindery.

30. Fine-silty over fragmental.
31. Fine-silty over sandy or sandy-skeletal.
32. Fine-silty over clayey if there is an absolute difference of >25 percent in the percentages of clay.
33. Clayey over fragmental.
34. Clayey over sandy or sandy-skeletal.
35. Clayey over loamy-skeletal if there is an absolute difference of >25 percent in the percentages of clay in the fine-earth fraction.
36. Clayey over loamy if there is an absolute difference of >25 percent in the percentages of clay.
37. Clayey over fine-silty if there is an absolute difference of >25 percent in the percentages of clay.
38. Thixotropic over fragmental.
39. Thixotropic over sandy or sandy-skeletal.
40. Thixotropic over loamy-skeletal.
41. Thixotropic over loamy.
42. Cindery over medial.
43. Cindery over medial-skeletal.
44. Ashy over medial.

The intent in setting up classes of strongly contrasting particle sizes is to identify changes in pore-size distribution that seriously affect movement and retention of water and that have not been identified in higher categories. The list given is intended for use in grouping the soil series of the United States into families. It is not intended as a complete list. For example, fine sand over coarse sand is common in the Udipsamments of western Europe but is not known to be important in the United States.

Choice of 7 or 11 particle-size classes

Only the seven particle-size classes are used in lithic, arenic, and grossarenic subgroups and in shallow families.

In families of Ultisols not included in the preceding item, subclasses of loamy particle-size classes are used but not subclasses of the clayey classes.

If only a part of the control section of a soil in an andic or andeptic subgroup or other group where substitute terms are used is cindery, ashy, medial, or thixotropic, contrasting families are recognized, but only the seven particle-size classes are used. For example, we might use cindery over loamy but not cindery over fine-loamy.

Only two particle-size classes are used to separate families in Vertisols, fine if there is <60 percent clay and very-fine if there is 60 percent or more clay in the weighted average of the control section.

Mineralogy classes

The control section

Mineralogy classes are based on the approximate mineralogical composition of selected size fractions of the same segment of the soil (control section) that is used for application of particle-size classes.

Contrasting mineralogy modifiers

Contrasting mineralogy modifiers are not recognized except where substitutes for particle-size class modifiers have been used. In those soils there is an overlay of ash or cinders, or an upper medial or thixotropic layer, and the ashy, cindery, medial, or thixotropic layer extends at least 10 cm into the upper part of the control section. In identifying and naming the contrasting mineralogy modifiers in families of those soils, the seven particle-size classes are used to describe the lower part of the section. For example, a pair of contrasting layers is named as medial over loamy, mixed, not medial over coarse-loamy, mixed.

If there are layers of contrasting particle size in the control section, the mineralogy class of the upper part of the control section is definitive of the family mineralogy. For example, if there is fine-loamy material of mixed mineralogy over sandy material that is siliceous, the proper modifiers to describe the family are fine-loamy over sandy, mixed, not fine-loamy, mixed, over sandy, siliceous.

Key to mineralogy classes

All mineral soils, except Oxisols, are placed in the first mineralogy class of the key (table 1) that accommodates them although they may appear also to meet the requirements of other mineralogy classes. The correct mineralogy class for Oxisols is determined by using the key in table 3. These are keys, not complete definitions. Substitute terms connoting both particle size and mineralogy are based on combined texture, consistence, and mineralogy classes and are used to indicate important variations in Andaquepts, Andepts, andic, andaqueptic, and andeptic subgroups, in cryic great groups and cryic subgroups of Spodosols, and in cindery and ashy families of Aridisols and Entisols. Mineralogy classes are not named in Calciaquolls because the effect of the carbonates overshadows other differences in mineralogy, and they are not named in Quartzipsamments, which, by definition, are siliceous.

It is recognized that it is normally impossible to be certain of the percentages of the various kinds of clay minerals. Quantitative methods of identification are still subject to change. Although much progress has been made in the past few decades, an element of judgment enters into the estimation. All the evidence does not need to come from X-ray, surface, and DTA determinations. Other physical and chemical properties suggest the mineralogy of many clayey soils. Changes in volume, cation-exchange capacity, and the consistence are useful in estimating the nature of clay.

The description of clay mineralogy in naming families of clayey soils is based on the weighted average of the control section.

Calcareous and reaction classes

The presence or absence of carbonates and the reaction are treated together because they are so intimately related. A calcareous horizon cannot be strongly acid. Calcareous classes are applied to the section between a depth of 25 and 50 cm or between a depth of 25 cm and a lithic or paralithic contact that is below a depth of 25 but not 50 cm, or to some part of the soil above a lithic or paralithic contact that is shallower than 25 cm. Two classes, calcareous and noncalcareous, are used in selected taxa. The definitions follow.

Calcareous.--The fine-earth fraction effervesces in all parts with cold dilute HCl.

Noncalcareous.--The fine-earth fraction does not effervesce in all parts with cold dilute HCl. The term noncalcareous is not used as a part of a family name.

It should be noted that a soil that contains dolomite is calcareous and that effervescence of dolomite, when treated with cold dilute HCl, is slow.

Reaction classes are applied to the control section that is defined for particle-size classes. Three classes (acid, nonacid, and allic) are used in selected taxa. The definitions follow.

Acid.--The pH is <5.0 in 0.01 1 M $CaCl_2$ (2:1) throughout the control section (about 5.5 in H_2O, 1:1).

Nonacid.--The pH is 5.0 or more in 0.01 M $CaCl_2$ (2:1) in at least some part of the control section. The term nonacid is not used in the family name of calcareous soils.

Allic.--There is more than 2 cmol(+) of KCl extractable Al per kg soil (<2 mm fraction) in some 30 cm layer in the control section.

Acid and nonacid classes are used only in the names of families of Entisols and Aquepts; they are not used in sandy, sandy-skeletal, and fragmental families of these taxa, nor are they used in Sulfaquepts and Fragiaquepts, or in families that have carbonatic or gypsic mineralogy. The allic class is used only in names of families of Oxisols.

Calcareous classes are used if appropriate in the same taxa as reaction classes and, in addition, are used in families of Aquolls except for Calciaquolls and and for Aquolls that have an argillic horizon. Calcareous and reaction classes are not used in soils that have carbonatic or gypsic mineralogy. A soil that is calcareous is never acid. Calcareous therefore implies nonacid, and both names are not used because nonacid would be redundant. Similarly, noncalcareous would be redundant in acid families, and it is not used as part of the family name. If calcareous is used in a family name, calcareous is considered to be a subclass of mineralogy. It follows the mineralogy class name and is shown in parenthesis, for example: fine-loamy, mixed (calcareous), mesic Typic Haplaquolls.

Soil temperature classes

Soil temperature classes, as named and defined here, are used as family differentiae in all orders. The names are used as family modifiers unless the name of a higher taxon

carries the same limitation. Thus, frigid is implied in all
boric suborders and cryic great groups, and is redundant if
used in the name of a family.

The Celsius (centigrade) scale is the standard. Approximate
Fahrenheit equivalents are indicated parenthetically. It is
assumed that the temperature is that of a soil that is not
being irrigated.

For soils in which the difference is 5°C (9°F) or more between mean summer (June, July, and August in the northern
hemisphere) and mean winter (December, January, and
February in the northern hemisphere) soil temperature at a
depth of 50 cm or at a lithic or paralithic contact, whichever
is shallower, the classes, defined in terms of the mean annual soil temperature, are as follows:

Frigid.--Below 8°C (47°F).

Mesic.--From 8° to 15°C (47° to 59°F).

Thermic.--From 15° to 22°C (59° to 72°F).

Hyperthermic.--22°C (72°F) or higher.

For soils in which the difference is less than 5°C (9°F) between mean summer and mean winter soil temperature at a
depth of 50 cm or at a lithic or paralithic contact, whichever
is shallower, the classes, defined in terms of the mean annual soil temperature, are as follows:

Isofrigid.--Below 8°C (47°F).

Isomesic.--From 8° to 15°C (47° to 50°F).

Isothermic.--From 15° to 22°C (59° to 72°F).

Isohyperthermic.--22°C (72°F) or higher.

The appropriate limit between isofrigid and isomesic cannot
be tested in the United States and probably will need to be
revised.

Other characteristics

Several soil characteristics other than those already mentioned are needed in particular taxa to provide reasonable
groupings of series into families. Some of these seem to be
logical family criteria. Others probably should have been
used in higher categories, but the lack of information about
them makes it much safer to use them as family differentiae
at this time. These characteristics include depth of soil,
consistence, moisture equivalent, slope of soil, and permanent cracks.

Depth of soil

Classes of shallow and deep soils may be needed at the family level in all the orders of mineral soils. Some distinctions
in depth are made in great groups and in arenic, paralithic,

and lithic subgroups, but some other soils should also be grouped in families according to depth. Some soils have a paralithic contact over soft rock such as clay shale that is too compact for penetration by roots. The soil depth classes follow:

Micro. Less than 18 cm through diagnostic horizons. Used in cryic great groups but not in pergelic subgroups or in Entisols.

Shallow. Less than 50 cm to the upper boundary of a duripan or a petrocalcic horizon or to a lithic, paralithic or a petroferric contact. Used in lithic and petroferric subgroups of Oxisols and all great groups of Alfisols, Aridisols, Entisols, Inceptisols, Mollisols, Spodosols, and Ultisols, except pergelic subgroups of the cryic great groups and lithic subgroups. It is emphasized that the adjective "shallow" is not used in the family name of lithic subgroups of orders, other than Oxisols, because it would be redundant.

<u>Consistence</u>

Some cemented horizons, for example, a duripan, are differentiae in the classification in categories above the family. Others such as a cemented spodic horizon (ortstein) are not, but no single family should include soils that have a continuous, shallow, cemented horizon and soils that do not. In Spodosols, in particular, a cemented spodic horizon needs to be used as a family differentia. The following classes of consistence are defined for Spodosols.

Ortstein. All or part of the spodic horizon is at least weakly cemented, when moist, into a massive horizon that is present in more than half of each pedon.

Noncemented. The spodic horizon, when moist, is not cemented into a massive horizon in as much as half of each pedon.

Cementation of a small volume into shot or concretions does not constitute cementation to form a massive horizon. The name of a family of noncemented Spodosols normally does not have a modifier to imply lack of cementation. The name of a family of cemented Spodosols contains the modifier "ortstein."

A cemented calcic or gypsic horizon is not identified in the family name. Many calcic and some gypsic horizons are weakly cemented and some are indurated. The recognition of a petrocalcic or petrogypsic horizon is expected to meet most, if not all, the needs for recognition of cementation in those horizons. Taxa of these cemented soils are not named in the family category.

<u>Classes of coatings (on sands)</u>

Despite the emphasis given to particle-size classes in the taxonomy, variability remains in the sandy particle-size class, which takes in sands and loamy sands. Some sands are very clean, almost completely free of silt and clay. Others are mixed with appreciable amounts of finer grains. A moisture equivalent of 2 percent makes a reasonable division of the sands at the family level. Two classes of Quartzipsamments are defined in terms of their moisture equivalent.

Coated. The moisture equivalent is 2 percent or more.

Uncoated. The moisture equivalent is <2 percent. The
moisture retained at tension of 50 kPa may be substituted
for the moisture equivalent. Or, if moisture tension data are
not available, the silt plus clay is <5 percent.
The moisture equivalent for this distinction is the weighted
average for the control section, weighted for the thickness
of each horizon or layer.

<u>Slope or shape of soil</u>

Soils of aquic great groups normally have level or concave
surfaces. They are mainly in places where ground water
saturates the soil during some period of the year. A few,
however, are on the sides of slopes where water cannot
stand and are kept wet by more or less continuous precipitation and by seepage of water from higher areas. In a very
few, the hydrostatic pressure keeps the soil wet. No consis-
tent internal morphologic clues have yet been found that
distinguish these sloping aquic soils if the dissolved oxygen
content is low, but their recognition in the field from the
position of the soil in the landscape is generally easy. In
aquic great groups, particularly in Aquolls, Aquox, and
Aquults, use the shape of the soil as a family differentia.
For Aquolls and Aquults use classes of level and sloping as
these classes are defined in the Soil Survey Manual. For
Aquox use sloping in the names of families if slope is >8
percent. It may be necessary to use slope classes as family
differentiae in other orders, but they should not be used in
families of Aquods or Albaqualfs. Level is assumed in families of aquic soils if no slope modifier is used in the family
name.

<u>Classes of permanent cracks</u>

Hydraquents consolidate[6] after drainage and become Fluvaquents. In the process, they form polyhedrons, roughly 12
to 50 cm in diameter, depending on the *n* value and particle
size. The polyhedrons are separated by cracks that range in
width from 2 mm to >1 cm. The polyhedrons may shrink
and swell with changes in moisture content of the soil, but
the cracks are permanent and can persist for some hundreds
of years even though the soils are cultivated. The cracks
permit rapid movement of water through the soil either vertically or laterally. Yet the soils may have the same particle
size, mineralogy, and other family properties as soils that are
not cracked or that have cracks that open and close with the
seasons. The soils that have permanent cracks are so rare in
the United States that only a provisional definition of their
characteristics can be presented.

The modifier "cracked" is used only to designate families of
Fluvaquents. It means that there are continuous, permanent,
lateral and vertical cracks, at least 2 mm wide, spaced at
average lateral intervals of 50 cm or less. If this modifier is
not in the family name, permanent cracks are assumed to be
absent.

---

[6]The process is designated by a Dutch word that means "to ripen" because the
change resembles the change in consistence of cheese as water is removed.

Family Differentiae For Histosols

Most of the differentiae used to distinguish families of Histosols have been defined earlier, either because they are
differentiae in mineral soils as well as in Histosols; or because their definitions are essential for the classification of
some Histosols in categories higher than the family. The
differentiae that are not defined elsewhere are defined in
this section and the taxa in which they are used are enu-
merated.

The order in which family modifiers are placed in the technical family names of Histosols follows. The modifiers
chosen are those appropriate to the particular family.
>   Particle size
>   Mineralogy, including nature of limnic deposits
>   Reaction
>   Soil temperature regime
>   Soil depth

The differentiae are discussed in the remainder of this section.

Particle-size classes

Particle-size modifiers are used in family names of Histosols
only in terric subgroups. The terms used follow.
>   Fragmental
>   Loamy-skeletal or clayey-skeletal
>   Sandy or sandy-skeletal
>   Loamy
>   Clayey

The meaning of each of these terms is the same as that defined for particle-size classes of mineral soils. The proper
term is selected to describe the weighted average particle
size of the upper 30 cm of the mineral layer or that part of
the mineral layer that is within the control section,
whichever is thicker.

Mineralogy classes

Mineralogy classes of Histosols are of three kinds, according
to the nature of the subgroup or great group.

Ferrihumic.--Containing ferrihumic materials within the
control section (applied to Fibrists, Hemists, and Saprists,
except Sphagnofibrists and sphagnic subgroups of other
great groups). Bog iron is present in some Histosols or in
organic soil materials. It is called ferrihumic material. It
consists of authigenic deposits (formed in place) of hydrated
iron oxides mixed with varying kinds or amounts of organic
materials. The iron in some places is present in large cemented aggregates. In others it may be mostly dispersed and
soft. The colors normally are; shades of dark reddish
brown, commonly mixed with black, and the colors change
little on drying. The content of iron oxide ranges from 10
percent to >20 percent.

Ferrihumic material either is saturated with water for long
periods (>6 months) or is in an artificially drained soil. The

content of free iron oxide should exceed 10 percent (7 percent Fe), but the horizon may be either organic or mineral provided there is at least 1 percent organic matter. The materials should have >2 percent (by weight) concretions of iron, which may range in size from fine (<5 mm) to 1 m or more in the largest lateral dimension. Colors should be dark reddish brown or reddish brown, or should be close to those colors. The presence of ferrihumic material within the control section is one of the family differentiae.

If ferrihumic is used as a modifier in the technical family name, no other mineralogy modifier is used because the presence of the iron is considered to be, by far, the most important characteristic.

<u>Modifiers applied only to terric subgroups.</u>--The mineralogy modifiers used for mineral soils are applied to the mineral parts of the soil for which a particle-size modifier has been used if the mineralogy is not ferrihumic.

<u>Modifiers applied to limnic subgroups.</u>--If limnic materials are present in the control section, if they are 5 cm or more thick, and if the materials do not have ferrihumic mineralogy, the following modifiers are used. These terms are defined in chapter 2.
Coprogenous. Limnic materials that consist of coprogenous earth are present.
Diatomaceous. Limnic materials that consist of diatomaceous earth are present.
Marly. Limnic materials that consist of marl are present.

Reaction classes

Modifiers to indicate reaction are used in all subgroups. The meanings follow.

<u>Euic.</u>--The pH of undried samples is 4.5 or more (0.01 M $CaCl_2$) in at least some part of the organic materials in the control section.

<u>Dysic.</u>--The pH is <4.5 (in 0.01 M $CaCl_2$) in all parts of the organic materials in the control section.

Soil temperature classes

Names and definitions of classes follow the rules given for soil temperature classes of mineral soils. Frigid, however, is redundant in boric and cryic great groups and is not used. No temperature modifier is used in pergelic subgroups.

Soil depth classes

Soil depth modifiers are used in all lithic subgroups of Histosols except in the suborder of Folists. It is assumed that lithic Folists have a shallow lithic contact. Other lithic Histosols have a lithic contact within the control section but it may be as much as 160 cm deep.

<u>Shallow families.</u>--Used in lithic subgroups to indicate a lithic contact between a depth of 18 cm and 50 cm.

<u>Micro families.</u>--Used to indicate a lithic contact shallower than 18 cm without regard to soil temperature. (In mineral soils, micro families are restricted to cryic great groups.)

LITERATURE CITED

Soil Survey Staff, 1951. Soil survey manual. U.S. Dept. Agr. Handb. 18. U.S. Govt. Printing Office, Washington, D.C.

Chapter 4

Identification of the Taxonomic
Class of a Soil

Key To Soil Orders

In this key and the other keys that follow, the diagnostic
horizons and the properties mentioned do not include the
properties of buried soils except their organic carbon if of
Holocene age and base saturation. Properties of buried soils
are considered in the categories of subgroups, families, and
series but not in those of order, suborder, and great group.
The meaning of the term "buried soil" has been given previously.

A. Soils that
either
1. Have organic soil materials that extend from the surface to one of the following:
    a. A depth within 10 cm or less of a lithic or par-
alithic contact, provided the thickness of the organic
soil materials is more than twice that of the mineral
soil above the contact; *or*
    b. Any depth if the organic soil material rests on
fragmental material (gravel, stones, cobbles) and the
interstices are filled with organic materials, or rests on
a lithic or paralithic contact;
or
2. Have organic materials that have an upper boundary
within 40 cm of the surface,
and
    a. Have one of the following thicknesses:
      (1) 60 cm or more if three-fourths or more of the
volume is moss fibers or the moist bulk density is
<0.1 g per cubic centimeter (6.25 lbs per cubic
foot); *or*
      (2) 40 cm or more if
        (a) The organic soil material is saturated with
water for long periods (>6 months) or is artificially drained; *and*
        (b) The organic material consists of sapric or
hemic materials or consists of fibric materials
that are less than three-fourths moss fibers by
volume and have a moist bulk density of 0.1 or
more;
and
    b. Have organic soil materials that
      (1) Do not have a mineral layer as much as 40 cm
thick either at the surface or whose upper boundary
is within a depth of 40 cm from the surface; *and*
      (2) Do not have mineral layers, taken cumulatively,
as thick as 40 cm within the upper 80 cm.
                      Histosols, p. 131

B. Other soils that do not have a plaggen epipedon but that have
either

   1. A spodic horizon whose upper boundary is within 2 m of the surface;

or

   2. A placic horizon that meets all the requirements of a spodic horizon except thickness and index of accumulation *and* rests on a fragipan, on a spodic horizon, or on an albic horizon that rests on a fragipan.

Spodosols, p. 227

C. Other soils that have:
either

   1. An oxic horizon with its upper boundary within 150 cm of the soil surface *and* do not have a clay content increase necessary to define the upper boundary of a kandic horizon within a depth of 150 cm of the soil surface,

or

   2. 40 percent or more clay in the surface 18 cm, after mixing, *and*, with its upper boundary within 150 cm of the soil surface, *either* an oxic horizon, *or* a kandic horizon that meets the weatherable mineral requirements of an oxic horizon.

Oxisols, p. 203

D. Other soils that

   1. Do not have a lithic or paralithic contact, petrocalcic horizon, or duripan within 50 cm of the surface;

and

   2. After the soil to a depth of 18 cm has been mixed, as by plowing, have 30 percent or more clay in all subhorizons to a depth of 50 cm or more;

and

   3. Have, at some time in most years unless irrigated or cultivated, open cracks[1] at a depth of 50 cm that are at least 1 cm wide and extend upward to the surface or to the base of the plow layer or surface crust

and

   4. Have *one or more* of the following;
      a. Gilgai;
      b. At some depth between 25 cm and 1 m, slickensides close enough to intersect; *or*
      c. At some depth between 25 cm and 1 m, wedge-shaped natural structural aggregates that have their long axes tilted 10° to 60° from the horizontal.

Vertisols, p. 259

---

[1]An open crack is interpreted to be a separation between gross polyhedrons. If the surface horizons are strongly self-mulching, that is, if the soil is a mass of loose granules, or if the soil is cultivated while the cracks are open, the cracks may be largely filled with granular materials from the surface. But they are considered to be open in the sense that the polyhedrons are separated.

E. Other soils that have an ochric or anthropic epipedon and *either*

    1. Do not have an argillic or a natric horizon but

        a. Are saturated with water within 1 m of the surface for 1 month or more in some years and have a salic horizon whose upper boundary is within 75 cm of the surface; *or*

        b. Have one or more of the following horizons whose upper boundary is within 1 m of the soil surface: a petrocalcic, calcic, gypsic, petrogypsic, or cambic horizon or a duripan; and have an aridic moisture regime,

or

    2. Have an argillic or a natric horizon and have

        a. An aridic moisture regime; *and*

        b. An epipedon that is not both massive and hard or very hard when dry.

Aridisols, p. 95

KEY

F. Other soils

- that have a mesic, isomesic, or warmer temperature regime, *and*

- do not have tongues of albic materials in the argillic horizon that have vertical dimensions of as much as 50 cm if there is >10 percent weatherable minerals in the 20- to 200-micrometer fraction, *but*

- have *one of the following* combinations of characteristics: *either*

    1. Have an argillic or kandic horizon but not a fragipan and have base saturation (by sum of cations) of <35 percent at the following depths:

        a. If the argillic or kandic horizon has in some part a hue of 5YR or yellower, or a color value, moist, of 4 or more, or a color value, dry, that is more than 1 unit higher than the value, moist, the shallowest of the following:

            (1) 1.25 m below the upper boundary of the argillic or kandic horizon;

            (2) 1.8 m below the surface of the soil; *or*

            (3) Immediately above a lithic or paralithic contact;

        b. If the argillic or kandic horizon has some other color or if the epipedon has a sandy or sandy-skeletal particle-size class throughout, the deepest of 1.25 m below the upper boundary of the argillic horizon, 1.8 m below the surface of the soil, or immediately above a lithic or a paralithic contact if it is shallower;

or

    2. Have a fragipan that

        a. Meets all the requirements of an argillic or kandic horizon or has clay skins >1 mm thick in some part, or underlies an argillic or kandic horizon;

    and

        b. Has base saturation (by sum of cations) of <35 percent at a depth of 75 cm below the upper boundary of the fragipan or immediately above a lithic or paralithic contact, whichever is shallower.

Ultisols, p. 237

G. Other soils that
    1. Have either of the following:
        a. A mollic epipedon;
or
        b. A surface horizon that, after the soil to a depth of 18 cm is mixed, meets all requirements of a mollic epipedon except thickness, and, in addition, have an upper subhorizon >7.5 cm thick that is in an argillic, kandic or a natric horizon, that meets the requirements of a mollic epipedon with respect to color, content of organic carbon, base saturation, and structure but is separated from the surface horizon by an albic horizon;
and, in addition,
    2. Have base saturation of 50 percent or more (by NH_4OAc) as follows:
either
        a. If there is an argillic, kandic or natric horizon, from its upper boundary to a depth of 1.25 m below that boundary, *or* to a depth 1.8 m below the soil surface *or* to a lithic or paralithic contact, whichever is least;
or
        b. If there is no argillic, kandic, or natric horizon, in all subhorizons to a depth 1.8 m below the soil surface or to a lithic or paralithic contact, whichever is least;
and, furthermore
    3. If the exchange complex is dominated by amorphous materials, have, in some subhorizon within a depth of 35 cm or to a lithic or paralithic contact shallower than 35 cm, a bulk density (at 33 kPa water tension) of the fine-earth fraction of 0.85 or more and have <60 percent vitric volcanic ash[2], cinders, or other pyroclastic material in the silt, sand, and gravel fractions within this depth;

Mollisols, p. 171

H. Other soils that
either
    1. Have an argillic, kandic or natric horizon but no fragipan;
or
    2. Have a fragipan that
        a. Is in or underlies an argillic or kandic horizon;
        or
        b. Meets all requirements of an argillic or kandic horizon; *or*
        c. Has clay skins >1 mm thick in some part.

Alfisols, p. 63

---

[2]Vitric material, as used here, includes glass and also crystalline particles that are coated with glass or with partly devitrified glass.

61

I. Other soils that
- have no sulfidic material within 50 cm of the mineral soil surface; *and*
- have between 20 and 50 cm below the mineral soil surface an *n* value of 0.7 or less in one or more subhorizons *or* less than 8 percent clay in one or more subhorizons; *and*
- have *one or more* of the following:
> 1. An umbric, mollic, histic (either mineral or organic) or plaggen epipedon; *or*
> 2. A cambic horizon or both an aquic moisture regime and permafrost; *or*
> 3. Within 1 m of the surface, a calcic, petrocalcic, gypsic, petrogypsic, or placic horizon or a duripan; *or*
> 4. A fragipan or an oxic horizon with its upper boundary between a depth of 150 and 200 cm; *or*
> 5. A sulfuric horizon whose upper boundary is within 50 cm of the soil surface; *or*
> 6. In half or more of the upper 50 cm, an SAR of $>13^3$ (or sodium saturation that is >15 percent) that decreases with depth below 50 cm and, within a depth of 1 m, have ground water at some period during the year when the soil is not frozen in any part.

Inceptisols, p. 145

J. Other soils

Entisols, p. 111

---

[3]The percentage of exchangeable sodium (ESP) is used in the definition of the natric horizon and in a number of the taxa. Since this text was written, the U.S. Salinity Laboratory (personal communication from C.A. Bower) has revised its definition of sodic (alkali) soils and the method for measuring the sodium adsorption ratio (SAR) as follows: SAR is measured by the normal method if the conductivity (EC) of the saturation extract is 20 dS m^{-1} at 25°C. If the conductivity is $\geq$ 20 dS m^{-1} and SAR is 10, SAR is determined on a sample that has been leached with distilled water until EC of the leachate decreases to about 4 dS m^{-1} but not to 4. ESP of $\geq$ 15 is replaced by SAR of $\geq$13 if EC is large enough to require a correction for soluble salts in calculating ESP. If EC is low enough ($\leq$4) that no correction is needed for soluble salts, ESP is determined directly from the replaced cations.

Chapter 5

Alfisols

Key To Suborders

HA. Alfisols that have an aquic moisture regime or are artificially drained and that have characteristics associated with wetness, namely, mottles, or iron-manganese concretions >2 mm in diameter, or chroma of 2 or less immediately below any Ap horizon or below any dark A horizon in which the moist color value is less than 3.5 after the material is rubbed, and one of the following:
    1. Dominant chroma of 2 or less[1] in coatings on the surface of peds and mottles within peds of the argillic or kandic horizon, or a dominant chroma of 2 or less in the matrix of the argillic or kandic horizon and mottles of higher chroma;
    2. If there are no mottles in the argillic or kandic horizon, a dominant chroma of 1 or less.

Aqualfs, p. 63

HB. Other Alfisols that have
    1. A frigid temperature regime but do not have a xeric moisture regime; or
    2. A cryic temperature regime.

Boralfs, p. 69

HC. Other Alfisols that have one of the following:
    1. An ustic moisture regime;
    2. An epipedon that is both massive and hard or very hard when dry and a moisture regime that is aridic but marginal to ustic;

Ustalfs, p. 81

HD. Other Alfisols that have one of the following:
    1. A xeric moisture regime; or
    2. An epipedon that is both massive and hard or very hard when dry and a moisture regime that is aridic but marginal to xeric.

Xeralfs, p. 88

HE. Other Alfisols that have a udic moisture regime.

Udalfs, p. 72

Aqualfs

Key to great groups

HAA. Aqualfs that have plinthite that forms a continuous phase or constitutes half or more of the matrix within some

---

[1] If the hue is redder than 10YR because of red parent materials that remain red after citrate-dithionite extraction, the requirement for low chroma is waived. Where the soil temperature regime is hyperthermic, isothermic, or isohyperthermic, chroma up to 4 is tentatively permitted if the hue is 2.5Y or 5Y and if mottles are distinct or prominent. Such soils are too few in the United States to permit testing these limits.

subhorizon between 30 cm and 150 cm below the surface of the soil.

Plinthaqualfs, p. 69

HAB. Other Aqualfs that have a natric horizon and do not have a duripan.

Natraqualfs, p. 67

HAC. Other Aqualfs that have a duripan.

Duraqualfs, p. 65

HAD. Other Aqualfs that have a fragipan.

Fragiaqualfs, p. 65

HAE. Other Aqualfs that have a CEC <16 cmol(+) kg^{-1} clay (by 1 M NH$_4$OAc pH7) and an ECEC <12 cmol(+) kg^{-1} clay (sum of bases extracted with 1 M NH$_4$OAc pH7 plus 1 M KCl-extractable Al) in the major part of the argillic or kandic horizon or the major part of the upper 100 cm of the argillic or kandic horizon if these horizons are thicker than 100 cm.

Kandiaqualfs, p. 66

HAF. Other Aqualfs that have an albic horizon tonguing into an argillic horizon.

Glossaqualfs, p. 66

HAG. Other Aqualfs that have an abrupt textural change between an ochric epipedon or an albic horizon and an argillic horizon and have slow or very slow hydraulic conductivity in the argillic horizon[2].

Albaqualfs, p. 64

HAH. Other Aqualfs that have an umbric epipedon.

Umbraqualfs, p. 69

HAI. Other Aqualfs.

Ochraqualfs, p. 67

Albaqualfs

Distinctions between Typic Albaqualfs and other subgroups

Typic Albaqualfs are the Albaqualfs that
a. Have chroma of 2 or less in 60 percent or more of the mass between the bottom of the A or the Ap horizon and a depth of 75 cm;
b. Do not have a layer above a depth of 75 cm that has texture finer than loamy fine sand, that is as much as 18 cm thick, that has a bulk density (at 33 kPa water tension) of 0.95 g per cubic centimeter or less in the fine-earth fraction, and that has either of the following:
  (1) A ratio of measured clay to 1500 kPa water
  (percentages) of 1.25 or less, or

---

[2]Hydraulic conductivity is defined as the rate of internal water movement under a unit potential gradient. In this text the term refers to vertical saturated hydraulic conductivity. Slow and very slow rates refer to 4 to 10 and less than 4 cm water per day respectively.

(2) A ratio of CEC (at pH near 8) to 1500 kPa water of
>1.5 and more exchange acidity than the sum of bases
plus KCl-extractable aluminum;
c. Do not have a horizon within a depth of 1 m from the
surface that is brittle, that is 15 cm or more thick, and that
contains some opal coatings or some (<20 percent) durinodes;
d. Either have an Ap horizon that has a color value, moist,
of 4 or more, or a color value, dry, of 6 or more after the
soil has been crushed and smoothed, or have, after the soil
to a depth of 18 cm has been mixed, an upper layer that has
these colors;
e. Do not have the following combination of characteristics:
  (1) Cracks at some period in most years that are 1 cm or
  more wide at a depth of 50 cm and at least 30 cm long in
  some part and that extend upward to the surface or to
  the base of an Ap or an albic horizon;
  (2) A coefficient of linear extensibility (COLE) of 0.09
  or more in a horizon or horizons at least 50 cm thick and
  a potential linear extensibility of 6 cm or more in the soil
  to a depth of 1 m or in the whole soil if the depth to a
  lithic or paralithic contact is >50 cm but <1 m;
  (3) More than 35 percent clay in horizons that have a
  total thickness of >50 cm;
f. Have texture that is very fine sand or finer in some subhorizon within a depth of 50 cm from the soil surface; and
g. Have a surface horizon that, after the soil to a depth of
18 cm has been mixed, has <30 percent clay and is continuous throughout each pedon.
Aeric Albaqualfs are like Typic Albaqualfs except for *a*.
Arenic Albaqualfs are like Typic Albaqualfs except for *f*,
with or without *a* or *d*, or both.
Mollic Albaqualfs are like Typic Albaqualfs except for *d*.
Ruptic-Vertic Albaqualfs are like Typic Albaqualfs except
for *e* and *g* with or without *a* or *d*, or both.
Udollic Albaqualfs are like Typic Albaqualfs except for *a*
and *d*.
Vertic Albaqualfs are like Typic Albaqualfs except for *e*
with or without *a* or *d*, or both.

Duraqualfs

Duraqualfs are the Aqualfs that have a duripan below the
argillic horizon. They are not known to occur in the United
States. The group has been proposed for other countries,
but definitions of subgroups have not been suggested.

Fragiaqualfs

Distinctions between Typic Fragiaqualfs and other subgroups

Typic Fragiaqualfs are the Fragiaqualfs that
a. Do not have a mottled horizon between the A or Ap
horizon and a fragipan that has dominant chroma more than
2 if the hue is 10YR or redder or more than 3 if the hue is
2.5Y or yellower;
b. Have <5 percent plinthite (by volume) in all subhorizons
within 1.5 m (60 in.) of the surface;
c. Have an Ap horizon that has either a color value, moist,
of 4 or more or a color value, dry, of 6 or more after the
soil has been crushed and smoothed; or the upper soil to a
depth of 18 cm, after mixing, has these color values;

Aeric Fragiaqualfs are like Typic Fragiaqualfs except for *a*.
Umbric Fragiaqualfs are like Typic Fragiaqualfs except for
c.

Glossaqualfs

Distinctions between Typic Glossaqualfs and other subgroups

Typic Glossaqualfs are the Glossaqualfs that
a. Have in 60 percent or more of the matrix[3] in all subhorizons between the A or Ap horizon and a depth of 75 cm
one of the following:
   (1) If mottled and the value, moist, is 4 or more, the
   chroma, moist, is 2 or less;
   (2) If not mottled, the chroma, moist, is 1 or less;
b. Have texture finer than loamy fine sand in some subhorizon within a depth of 50 cm below the surface; and
c. Have an Ap horizon that has either a color value, moist,
of 4 or more, or a color value, dry, of 6 or more after the
soil has been crushed and smoothed; or the soil to a depth of
18 cm, after mixing, has these colors.
Aeric Glossaqualfs are like Typic Glossaqualfs except for *a*
or for *a* and *c*.
Arenic Glossaqualfs are like Typic Glossaqualfs except for *b*
or for *a* and *b* ; they have a sandy epipedon between 50 cm
and 1 m thick.
Mollic Glossaqualfs are like Typic Glossaqualfs except for *c*.

Kandiaqualfs

Distinctions between Typic Kandiaqualfs and other subgroups

Typic Kandiaqualfs are the Kandiaqualfs that
a. Have in 60 percent or more of the matrix in all subhorizons between the A or Ap horizon and a depth of 75 cm
one or more of the following:
   (1) If mottled and the mean annual soil temperature is
   lower than 15°C, has chroma, moist, of 2 or less;
   (2) If mottled and the mean annual soil temperature is
   15°C or more:
      (a) If the hue is 2.5Y or redder and the value, moist,
      is more than 5, the chroma, moist, is 2 or less;
      (b) If the hue is 2.5Y or redder and the value, moist,
      is 5 or less, the chroma, moist, is 1 or less;
      (c) If the hue is yellower than 2.5Y, the chroma,
      moist, is 2 or less;
   (3) The chroma, moist, is 1 or less whether mottled or
   not;
b. Have an Ap horizon that has either a color value, moist,
of 4 or more or a color value, dry, or 6 or more after the
soil has been crushed; or the upper soil to a depth of 18 cm,
after mixing, has these color values;
c. Do not have an epipedon that is 50 cm to 100 cm thick if
the particle-size class is sandy throughout;

---

[3]If the hue is 7.5YR or redder and, if peds are present, ped exteriors in the
argillic horizon should have dominant chroma, moist, of 1 or less and ped interiors should have mottles that have chroma, moist, of 2 or less; if peds are
absent, the chroma, moist, should be 1 or less immediately below any surface
horizon that has color value, moist, less than 3.5.

d. Do not have an epipedon that is >100 cm thick if the particle-size class is sandy throughout: and
e. Do not have a horizon within 150 cm of the soil surface that has >5 percent plinthite by volume.
Aeric Kandiaqualfs are like Typic Kandiaqualfs except for *a*.
Aeric Umbric Kandiaqualfs are like Typic Kandiaqualfs except for *a* and *b*.
Arenic Kandiaqualfs are like Typic Kandiaqualfs except for *c*, with or without *a*.
Grossarenic Kandiaqualfs are like Typic Kandiaqualfs except for *d*, with or without *a*.
Plinthic Kandiaqualfs are like Typic Kandiaqualfs except for *e*.
Umbric Kandiaqualfs are like Typic Kandiaqualfs except for *b*.

Natraqualfs

Distinctions between Typic Natraqualfs and other subgroups

Typic Natraqualfs are Natraqualfs that
a. Have >15 percent saturation with sodium or have more magnesium and sodium than calcium and extractable acidity within 15 cm of the upper boundary of the natric horizon;
b. Do not have tonguing or interfingering of albic materials more than 2.5 cm into the natric horizon;
c. Have an Ap horizon that has either a color value, moist, of 4 or more or a color value, dry, of 6 or more after the soil has been crushed and smoothed; or the soil to a depth of 18 cm, after mixing, has these colors; and
d. Have, within 40 cm of the soil surface, a horizon that has 15 percent or more saturation with sodium or has more magnesium and sodium than calcium and extractable acidity.
Albic Natraqualfs are like Typic Natraqualfs except for *a* or for *a* and *d*.
Albic Glossic Natraqualfs are like Typic Natraqualfs except for *b* and *d*, with or without *a*.
Glossic Natraqualfs are like Typic Natraqualfs except for *b*.
Mollic Natraqualfs are like Typic Natraqualfs except for *c*, with or without *d*.

Ochraqualfs

Distinctions between Typic Ochraqualfs and other subgroups

Typic Ochraqualfs are the Ochraqualfs that
a. Have in 60 percent or more of the matrix[4] in all subhorizons between the A or Ap horizon and a depth of 75 cm one or more of the following:
    (1) If mottled and the mean annual soil temperature is lower than 15°C, chroma, moist, of 2 or less;
    (2) If mottled and the mean annual soil temperature is 15°C or more:

---

[4]If the hue is 7.5YR or redder and, if peds are present, ped exteriors in the argillic horizon should have dominant chroma, moist, of 1 or less and ped interiors should have mottles that have chroma, moist, of 2 or less; if peds are absent, the chroma, moist, should be 1 or less immediately below any surface horizon that has color value, moist, less than 3.5.

(a) If the hue is 2.5Y or redder and the value, moist, is more than 5, the chroma is 2 or less;

(b) If the hue is 2.5Y or redder and the value, moist, is 5 or less, the chroma, moist, is 1 or less;

(c) If the hue is yellower than 2.5Y, the chroma, moist is 2 or less;

(3) The chroma, moist, is 1 or less whether mottled or not;

b. Do not have a layer above a depth of 75 cm that has a texture finer than loamy fine sand, that is as much as 18 cm thick, that has bulk density (at 33 kPa water tension) of 0.95 g per cubic centimeter or less in the fine-earth fraction, and that has either

(1) A ratio of measured clay to 1500 kPa water (percentages) of 1.25 or less; or

(2) A ratio of CEC (at pH near 8) to 1500 kPa water of >1.5 and more exchange acidity than the sum of bases plus KCl-extractable aluminum;

c. Have an Ap horizon that has either a color value, moist, of 4 or more or a color value, dry, of 6 or more after the soil has been crushed and smoothed; or the upper soil to a depth of 18 cm, after mixing, has these color values;

d. Have texture finer than loamy fine sand in some sub-horizon within 50 cm of the surface; and

e. Do not have the following combination of characteristics:

(1) Cracks at some period in most years that are 1 cm or more wide at a depth of 50 cm and at least 30 cm long in some part and that extend upward to the surface or to the base of an Ap or an albic horizon,

(2) A coefficient of linear extensibility (COLE) of 0.09 or more in a horizon or horizons at least 50 cm thick and a potential linear extensibility of 6 cm or more in the upper soil to a depth of 1 m or in the whole soil if a lithic, paralithic, or petroferric contact is deeper than 50 cm but not deeper than 1 m, and

(3) More than 35 percent clay in horizons that have total thickness of >50 cm.

Aeric Ochraqualfs are like Typic Ochraqualfs except for a.

Aeric Umbric Ochraqualfs are like Typic Ochraqualfs except for *a* and *c* and have an epipedon that meets all the requirements of an umbric epipedon except thickness.

Andaqueptic Ochraqualfs are like Typic Ochraqualfs except for *b* or for *a* and *b*.

Arenic Ochraqualfs are like Typic Ochraqualfs except for *d*, with or without *a*, or *c*, or both, and have a sandy epipedon between 50 cm and 1 m thick.

Grossarenic Ochraqualfs are like Typic Ochraqualfs except for *d*, or for *a* and *d*, and have a sandy epipedon >1 m thick.

Mollic Ochraqualfs are like Typic Ochraqualfs except for *c* and have an epipedon that meets all the requirements of a mollic epipedon except thickness.

Udollic Ochraqualfs are like Typic Ochraqualfs except for *a* and *c* and have an epipedon that meets all the requirements of a mollic epipedon except thickness.

Umbric Ochraqualfs are like Typic Ochraqualfs except for *c* and have an epipedon that meets all the requirements of a umbric epipedon except thickness.

Vertic Ochraqualfs are like Typic Ochraqualfs except for *e* with or without *a* or *c*, or both.

Plinthaqualfs

Plinthaqualfs are the Aqualfs that have plinthite that forms
a continuous phase or that constitutes half or more of the
matrix of some subhorizon of the argillic horizon within 150
cm of the soil surface.

Umbraqualfs

Distinctions between Typic Umbraqualfs and other subgroups

Typic Umbraqualfs are the Umbraqualfs that
a. Do not have a layer above a depth of 75 cm that has a
texture finer than loamy fine sand, that is as much as 18 cm
thick, that has a bulk density (at 33 kPa water tension) of
0.95 g per cubic centimeter or less in the fine-earth fraction, and that has either
    (1) a ratio of measured clay to 1500 kPa water
    (percentages) of 1.25 or less, or
    (2) a ratio of CEC (at pH near 8) to 1500 kPa water of
    >1.5 and more exchange acidity than the sum of bases
    plus KCl-extractable aluminum; and
b. Do not have, in the umbric epipedon and in horizons
above the argillic horizon, soft discrete nodules 2.5 to 30 cm
in diameter that constitute >5 percent of the volume, that
are cemented by iron, and that lie above and in an irregular
or broken upper boundary of the argillic horizon.

The typic subgroup is the only one that has been recognized
in the United States to date. Item a of the definition of
Typic Umbraqualfs provides for an andeptic subgroup, and
item b has been suggested for a ferrudalfic subgroup.

The Umbraqualfs are not extensive in the United States.
Most of them formed in alluvium or in marine deposits and
are nearly level. The base saturation in the argillic horizon
generally is low enough that liming the surface layer does
not change the classification.

Boralfs

Key to great groups

HBA. Boralfs that have an argillic horizon with its upper
boundary deeper than 60 cm below the mineral surface[5],
that have texture finer than loamy fine sand in some subhorizon above the argillic horizon, and that have albic materials tonguing or interfingering in the argillic horizon.

Paleboralfs, p. 72

HBB. Other Boralfs that have a fragipan.

Fragiboralfs, p. 71

HBC. Other Boralfs that have a natric horizon.

Natriboralfs, p. 72

---

[5]If there is a surface mantle that has > 60 percent vitric volcanic ash, cinders,
or other vitric pyroclastic materials, the depth to the argillic horizon is measured from the base of this mantle rather than from the surface of the mineral
soil.

HBD. Other Boralfs that have a cryic temperature regime.

Cryoboralfs, p. 70

HBE. Other Boralfs that have base saturation (by sum of cations) of 60 percent or more in all subhorizons of the argillic horizon and are dry in some horizon at some time in most years.

Eutroboralfs, p. 70

HBF. Other Boralfs that either are never dry in any horizon in most years or have base saturation (by sum of cations) of <60 percent in some subhorizon of the argillic horizon.

Glossoboralfs, p. 71

Cryoboralfs

Distinctions between Typic Cryoboralfs and other subgroups

Typic Cryoboralfs are the Cryoboralfs that
a. Do not have a layer above a depth of 75 cm that has texture finer than loamy fine sand, that is as much as 18 cm thick, that has a bulk density (at 33 kPa water tension) of 0.95 g per cubic centimeter or less in the fine-earth fraction, and that has either (1) a ratio of measured clay to 1500 kPa water (percentages) of 1.25 or less or (2) a ratio of CEC (at pH near 8) to 1500 kPa water >1.5 and has more exchange acidity than the sum of bases plus KCl-extractable aluminum;
b. Do not have albic materials tonguing in an argillic horizon;
c. Do not have a lithic contact within 50 cm of the surface;
d. Have an Ap horizon that has a color value, moist, of more than 3, or the upper soil, to a depth of 15 cm after mixing, has a moist color value of 4 or more;
e. Have an argillic horizon that has a texture finer than loamy fine sand and is continuous vertically for at least the upper 15 cm (not in lamellae); and
f. Do not have mottles that have chroma of 2 or less within 75 cm of the surface, or the soils are not continuously saturated with water for as long as 3 months within 1 m of the surface where undrained.
Andeptic Cryoboralfs are like Typic Cryoboralfs except for *a* or for *a* and *b*.
Aquic Cryoboralfs are like Typic Cryoboralfs except for *f*, with or without *b* or *d*, or both.
Glossic Cryoboralfs are like Typic Cryoboralfs except for *b*.
Lithic Cryoboralfs are like Typic Cryoboralfs except for *c*.
Lithic Mollic Cryoboralfs are like Typic Cryoboralfs except for *c* and *d*, with or without *b*.
Mollic Cryoboralfs are like Typic Cryoboralfs except for *d* or *b* and *d*.
Psammentic Cryoboralfs are like Typic Cryoboralfs except for *e*, with or without *b* or *d*, or both.

Eutroboralfs

Distinctions between Typic Eutroboralfs and other subgroups

Typic Eutroboralfs are the Eutroboralfs that
a. Have an argillic horizon that

(1) If its upper boundary is <50 cm below the soil surface, does not have mottles that have chroma of 2 or less in the upper 25 cm if it is saturated with water within that depth at some time when the soil temperature is 5°C or higher; or
(2) If the upper boundary of the argillic horizon is 50 cm or more, does not have mottles that have chroma of 2 or less within a depth of 75 cm below the soil surface;
b. Have a texture finer than loamy fine sand in some subhorizon within 50 cm of the surface;
c. Do not have tongues of albic materials in the argillic horizon (interfingering is permitted);
d. Do not have a lithic contact within 50 cm of the surface;
e. Have an Ap horizon that has a color value, moist, of 4 or more or a color value, dry, of 6 or more (crushed and smoothed), or the soil to a depth of 18 cm, after mixing, has these colors; and
f. Have an argillic horizon that has a texture finer than loamy fine sand and is continuous vertically for at least the upper 15 cm (not in lamellae).
Aquic Eutroboralfs are like Typic Eutroboralfs except for *a*, with or without *e* or *f* or both.
Aquic Arenic Eutroboralfs are like Typic Eutroboralfs except for *a* and *b* with or without *c* or *e* or both.
Arenic Eutroboralfs are like Typic Eutroboralfs except for *b*, with or without *c* or *e*, or both.
Glossaquic Eutroboralfs are like Typic Eutroboralfs except for *a* and *c* with or without *e*.
Glossic Eutroboralfs are like Typic Eutroboralfs except for *c*.
Lithic Eutroboralfs are like Typic Eutroboralfs except for *d*, with or without all or any of *a*, *b*, *e*, or *f*.
Mollic Eutroboralfs are like Typic Eutroboralfs except for *e*.
Psammentic Eutroboralfs are like Typic Eutroboralfs except for *f* with or without all or any of *b*, *c*, or *e*.

Fragiboralfs

Distinctions between Typic Fragiboralfs and other subgroups

Typic Fragiboralfs are the Fragiboralfs that
a. Do not have mottles that have chroma of 2 or less (defined in ch. 1) in the upper 25 cm of the argillic horizon (bleached coatings of silt or sand may be on peds beside or below tongues of the albic horizon).
Aquic Fragiboralfs are like Typic Fragiboralfs except they have mottles that have chroma of 2 or less within the upper 25 cm of the argillic horizon and are saturated with water at some time within this depth.

Glossoboralfs

Distinctions between Typic Glossoboralfs and other subgroups

Typic Glossoboralfs are the Glossoboralfs that
a. Do not have mottles that have chroma of 2 or less in the upper 25 cm of the argillic horizon if the mottled horizons are saturated with water at a time when the soil temperature is 5°C or higher (bleached coatings of silt or sand may be on peds beside or below tongues of the albic horizon);

b. Have tongues of albic materials in the argillic horizon;
c. Do not have a lithic contact within 50 cm of the soil surface; and
d. Have an argillic horizon that has a texture finer than loamy fine sand and is continuous vertically for at least the upper 15 cm (not in lamellae).
Aquic Glossoboralfs are like Typic Glossoboralfs except for *a*, with or without *b* or *d*, or both.
Eutric Glossoboralfs are like Typic Glossoboralfs except for *b*.
Lithic Glossoboralfs are like Typic Glossoboralfs except for *c*, with or without *a* or *b*, or both.
Psammentic Glossoboralfs are like Typic Glossoboralfs except for *d*, with or without *b*.

Natriboralfs

Natriboralfs are the Boralfs that have a natric horizon. They are rare in the United States, and subgroups have not been developed.

Paleboralfs

Distinctions between Typic Paleboralfs and other subgroups

Typic Paleboralfs are the Paleboralfs that
a. Have an argillic horizon that has an increase in clay content of <20 percent (absolute) within a vertical distance of 7.5 cm from its upper boundary;
b. Do not have a layer in the upper 75 cm that has a texture finer than loamy fine sand, that is as much as 18 cm thick, that has a bulk density (at 33 kPa water tension) of 0.95 g per cubic centimeter or less in the fine-earth fraction, and that has either
    (1) a ratio of measured clay to 1500 kPa water (percentages) of 1.25 or less or
    (2) a ratio of CEC (at pH near 8) to 1500 kPa water of >1.5 and more exchange acidity than the sum of bases plus KCl-extractable aluminum;
c. Do not have mottles that have chroma of 2 or less within 1 m of the surface; and
d. Have an Ap horizon that has a color value, moist, of 4 or more or a color value, dry, of 6 or more when crushed and smoothed, or the soil to a depth of 18 cm has these colors after mixing.
Aquic Paleboralfs are like Typic Paleboralfs except for *c*.
Mollic Paleboralfs are like Typic Paleboralfs except for *d*.
The Paleboralfs have had relatively little study in the field or laboratory. It seems best at this time to consider the definitions of the great group and of its subgroups as tentative.

Udalfs

Key to great groups

HEA. Udalfs that have an agric horizon.
Agrudalfs, p. 74

HEB. Other Udalfs that have a natric horizon.
Natrudalfs, p. 79

HEC. Other Udalfs that
  1. Do not have a continuous albic horizon above the argillic horizon;
  2. Have a broken upper boundary of the argillic horizon; and
  3. Have discrete nodules in the argillic horizon that range from 2.5 to 5 cm to about 30 cm in diameter; exteriors of nodules are enriched and weakly cemented or indurated with iron and have redder hue or stronger chroma than interiors of nodules.

Ferrudalfs, p. 74

HED. Other Udalfs that have tongues of albic materials in the argillic horizon and do not have a fragipan.

Glossudalfs, p. 76

HEE. Other Udalfs that have tongues of albic materials in the argillic horizon and have a fragipan.

Fraglossudalfs, p. 75

HEF. Other Udalfs that have a fragipan.

Fragiudalfs, p. 74

HEG. Other Udalfs that
1. Have a CEC <16 cmol(+) kg^{-1} clay (by 1 M NH$_4$OAc pH 7) and an ECEC <12 cmol(+) kg^{-1} clay (sum of bases extracted with 1 M NH$_4$OAc pH 7 plus 1 M KCl-extractable Al) in the major part of the argillic or kandic horizon or the major part of the upper 100 cm of the argillic or kandic horizon if these horizons are thicker than 100 cm;
2. Do not have a lithic, paralithic, or petroferric contact within 150 cm of the soil surface; and
3. Have a clay distribution such that the percentage of clay does not decrease from its maximum amount by as much as 20 percent within a depth of 150 cm from the soil surface or the layer in which the clay percentage decreases by more than 20 percent has at least 5 percent of the volume consisting of skeletans on faces of peds and there is at least 3 percent (absolute) increase in clay below the layer.

Kandiudalfs, p. 78

HEH. Other Udalfs that have a CEC <16 cmol(+) kg^{-1} clay (by 1 M NH$_4$OAc pH 7) and an ECEC <12 cmol(+) kg^{-1} clay (sum of bases extracted with 1 M NH$_4$OAc pH 7 plus 1 M KCl-extractable Al) in the major part of the argillic or kandic horizon or the major part of the upper 100 cm of the argillic or kandic horizon if these horizons are thicker than 100 cm;

Kanhapludalfs, p. 78

HEI. Other Udalfs that
  1. Do not have a lithic or paralithic contact within 1.5 m of the soil surface;
  2. Have clay distribution such that the percentage of clay does not decrease by as much as 20 percent of the maximum within a depth of 150 cm from the soil surface or the layer in which the clay percentage decreases by more than 20 percent has at least 5 percent of the volume consisting of skeletans on faces of peds and there is at least 3 percent (absolute) increase in clay below the layer; and

3. Have one or more of the following in the argillic horizon:
    a. Hue redder than 10YR and chroma more than 4 dominant in the matrix in at least the lower part;
    b. Hue of 2.5YR or redder and value, moist, of less than 4 and value, dry, of less than 5 throughout the major part;
    c. Many coarse mottles that have hue redder than 7.5YR or chroma more than 5, or both in some sub-horizon.

Paleudalfs, p. 79

HEJ. Other Udalfs that have an argillic horizon that has throughout its thickness a hue redder than 5YR, a color value, moist of less than 3.5, and a color value, dry, no more than 1 unit higher than the value, moist.

Rhodudalfs, p. 80

HEK. Other Udalfs.

Hapludalfs, p. 76

Agrudalfs

Agrudalfs are the Udalfs that have an agric horizon. Some but not all have an anthropic epipedon. They have been in farms for many hundreds of years and have received heavy applications of animal manure and other amendments. They are not known to occur in the United States. It seems probable that only the typic and anthropic subgroups are needed. The Agrudalfs of western Europe have been farmed for more than 1,000 years[6], and the early farmers selected only the best soils for cultivation. All of them are well drained, and they seem quite similar in most properties. The Typic Agrudalfs do not have an anthropic epipedon.

Ferrudalfs

Distinctions between Typic Ferrudalfs and other subgroups

Typic Ferrudalfs are the Ferrudalfs that
a. Do not have mottles that have chroma of 2 or less within the upper 60 cm if the horizons that have mottles of low chroma are saturated with water at some time of year or if the soil has artificial drainage. The mottles should be distinguished from skeletans that may also have low chroma.
Aquic Ferrudalfs are like Typic Ferrudalfs except for *a.*

Fragiudalfs

Distinctions between Typic Fragiudalfs and other subgroups

Typic Fragiudalfs are the Fragiudalfs that
a. Have an argillic horizon above the fragipan that has clay skins on at least some vertical and horizontal faces of primary or secondary peds, or both;
b. Do not have, immediately above the fragipan, thick skeletans of clean sand and silt on primary ped faces and do not have an eluvial horizon (E') that has thick skeletans and

---

[6]Personal communication from R. Tavernier.

as much as 3 percent (absolute) less clay than both the
overlying and underlying horizons;
c. Do not have a layer in the upper 75 cm that has a texture
finer than loamy fine sand, that is as much as 18 cm thick,
that has a bulk density (at 33 kPa water tension) of 0.95 g
per cubic centimeter or less in the fine-earth fraction, and
that has either
    (1) a ratio of measured clay to 1500 kPa water
    (percentages) of 1.25 or less or
    (2) a ratio of CEC (at pH near 8) to 1500 kPa water of
    >1.5 and more exchange acidity than the sum of bases
    plus KCl-extractable aluminum;
d. Either have an Ap horizon that has a color value, moist,
of 4 or more or a color value, dry, of 6 or more, when
crushed and smoothed, or the soil to a depth of 18 cm, after
mixing, has those colors; and
e. Do not have mottles that have chroma of 2 or less in the
upper 25 cm of the argillic horizon and do not have mottles
that have chroma of 2 or less within 40 cm of the surface if
the horizons that have mottles of low chroma are saturated
with water at some time of year when the soil temperature
is 5°C or higher in those horizons. Mottles are not the same
as skeletans that may also have low chroma.

Albaquic Fragiudalfs are like Typic Fragiudalfs except for
e, and they have, within a vertical distance of 7.5 cm at the
top of the argillic horizon, a clay increase of >15 percent
(absolute) in the fine-earth fraction.

Aqueptic Fragiudalfs are like Typic Fragiudalfs except for *a*
and *e*.

Aquic Fragiudalfs are like Typic Fragiudalfs except for *e*.

Glossaquic Fragiudalfs are like Typic Fragiudalfs except for
b and *e*, with or without *a*.

Glossic Fragiudalfs are like Typic Fragiudalfs except for *b*,
with or without *a*.

Mollic Fragiudalfs are like Typic Fragiudalfs except for *d*,
with or without *e*, and have base saturation of 50 percent or
more (by NH_4OAc) in the major part of the epipedon.

Ochreptic Fragiudalfs are like Typic Fragiudalfs except for
a.

Umbreptic Fragiudalfs are like Typic Fragiudalfs except for
a, *b*, and *d*, and have base saturation <50 percent (by
NH_4OAc) in the epipedon.

Fraglossudalfs

Fraglossudalfs have a fragipan and an overlying argillic
horizon that has evidences of destruction in the form of
deep wide tongues of albic materials that normally extend
through the argillic horizon. The upper boundary of the
argillic horizon is usually broken, but there is little or no
cementation by iron. These soils are more extensive in Europe than in the United States. The epipedon may approach
an umbric epipedon.

Distinctions between Typic Fraglossudalfs and other subgroups

Typic Fraglossudalfs are the Fraglossudalfs that
a. Do not have mottles that have a chroma of 2 or less in
the upper 25 cm of the argillic horizon if the mottled hori-

zons are saturated with water at some season when the soil temperature is 5°C.

Aquic Fraglossudalfs are like Typic Fraglossudalfs except for *a*.

Glossudalfs

Distinctions between Typic Glossudalfs and other subgroups

Typic Glossudalfs are the Glossudalfs that
a. Do not have mottles that have chroma of 2 or less in the upper 25 cm of the argillic horizon if the mottled horizons are saturated with water at some season when their temperature is 5°C or higher;
b. Have tongues of albic materials that extend through at least the upper 50 cm of the argillic horizon;
c. Do not have a brittle matrix in one-fourth or more of some subhorizon that is at least 10 cm thick and that has an upper boundary within 1.25 m of the surface; and
d. Have within 50 cm of the surface a texture finer than loamy fine sand in some subhorizon.

Aquic Glossudalfs are like Typic Glossudalfs except for *a*, with or without *b*.

Arenic Glossudalfs are like Typic Glossudalfs except for *d*.

Fragic Glossudalfs are like Typic Glossudalfs except for *c*, or for *a* and *c*.

Haplic Glossudalfs are like Typic Glossudalfs except for *b*.

Hapludalfs

Distinctions between Typic Hapludalfs and other subgroups

Typic Hapludalfs are the Hapludalfs that
a. Do not have an abrupt textural change if there are mottles in the upper 25 cm of the argillic horizon;
b. Do not have a layer in the upper 75 cm that has texture finer than loamy fine sand, that is as much as 18 cm thick, that has a bulk density (at 33 kPa water tension) of 0.95 g per cubic centimeter or less in the fine-earth fraction, and that has either of the following:
    (1) A ratio of measured clay to 1500 kPa water (percentages) of 1.25 or less; or
    (2) A ratio of CEC (at pH near 8) to 1500 kPa water of >1.5 and more exchange acidity than the sum of bases plus KCl-extractable aluminum;
c. Have an argillic horizon that
    (1) If its upper boundary is <50 cm below the soil surface, does not have mottles that have chroma of 2 or less in the upper 25 cm if it is saturated with water within that depth at some time when the soil temperature is 5°C or higher; or
    (2) If the upper boundary of the argillic horizon is deeper than 50 cm, does not have mottles that have chroma of 2 or less within a depth of 75 cm below the soil surface;
d. Have, within 50 cm of the surface, texture finer than loamy fine sand in some subhorizon;
e. Do not have interfingering of albic materials and albic materials surrounding some peds in the upper part of the argillic horizon;

f. Do not have a lithic contact within 50 cm of the soil surface;

g. Have an Ap horizon that has a color value, moist, of 4 or more or has a color value, dry, of 6 or more (crushed and smoothed) or the upper soil to a depth of 18 cm, after mixing, has these colors;

h. Have an argillic horizon that is continuous horizontally and continuous vertically for at least the upper 20 cm of its thickness and that has texture finer than loamy fine sand;

i. Have base saturation (by sum of cations) of 60 percent or more at a depth 1.25 m below the top of the argillic horizon, or 1.8 m below the soil surface, or immediately above a lithic or paralithic contact, whichever is least;

j. Do not have the following combinations of characteristics:

    (1) Cracks at some period in most years that are 1 cm or more wide at a depth of 50 cm and at least 30 cm long in some part and that extend to the surface or to the base of an Ap horizon;

    (2) A coefficient of linear extensibility (COLE) of 0.09 or more in a horizon or horizons at least 50 cm thick and a potential linear extensibility of 6 cm or more in the upper 1 m of the soil or in the whole soil if a lithic or paralithic contact is deeper than 50 cm but shallower than 1 m; and

    (3) More than 35 percent clay in horizons that total >50 cm in thickness within the control section; and

k. Do not have albic materials that constitute as much as 5 percent of the volume of any subhorizon of the argillic horizon.

Albaquic Hapludalfs are like Typic Hapludalfs except for *a* and *c*, with or without *g*.

Albaquultic Hapludalfs are like Typic Hapludalfs except for *a*, *c*, and *i*.

Andeptic Glossoboric Hapludalfs are like Typic Hapludalfs except for *b* and *e*, and their mean annual soil temperature is lower than 10°C.

Aquic Hapludalfs are like Typic Hapludalfs except for *c*.

Aquic Arenic Hapludalfs are like Typic Hapludalfs except for *c* and *d*, with or without *g* or *i*, or both.

Aquic Lithic Hapludalfs are like Typic Hapludalfs except for *c* and *f*.

Aquollic Hapludalfs are like Typic Hapludalfs except for *c* and *g*.

Aquultic Hapludalfs are like Typic Hapludalfs except for *c* and *i*, with or without *g*.

Arenic Hapludalfs are like Typic Hapludalfs except for *d*, with or without *g* or *i*, or both.

Glossaquic Hapludalfs are like Typic Hapludalfs except for *e* and *c* or for *e*, *c*, and *g*.

Glossic Hapludalfs are like Typic Hapludalfs except for *k*, with or without *e* or *g*, or both, and the mean annual soil temperature is 10°C or higher.

Glossoboric Hapludalfs are like Typic Hapludalfs except for *e* or for *e* and *g*, and their mean annual soil temperature is lower than 10°C.

Lithic Hapludalfs are like Typic Hapludalfs except for *f*, with or without *g*.

Mollic Hapludalfs are like Typic Hapludalfs except for *g*.

Psammaquentic Hapludalfs are like Typic Hapludalfs except for *c* and *h*, with or without all or any of *d*, *g*, or *i*.

Psammentic Hapludalfs are like Typic Hapludalfs except for *h*, with or without all or any of *d*, *g*, or *i*.
Ultic Hapludalfs are like Typic Hapludalfs except for *i*, with or without *g*.
Vertic Hapludalfs are like Typic Hapludalfs except for *j*, with or without any or all of *a*, *c*, *g*, and *i*.

Kandiudalfs

Distinctions between Typic Kandiudalfs and other subgroups

Typic Kandiudalfs are the Kandiudalfs that
a. Do not have mottles that have chroma of 2 or less within 75 cm of the soil surface if the mottled horizon is saturated with water at some time when the soil temperature at that depth is 5°C or higher or the soil has artificial drainage;
b. Do not have an epipedon that is 50 cm to 100 cm thick if the particle-size class is sandy throughout;
c. Do not have an epipedon that is >100 cm thick if the particle-size class is sandy throughout;
d. Have an Ap horizon that has either a color value, moist, of 4 or more or a color value, dry, or 6 or more after the soil has been crushed, or the upper soil to a depth of 18 cm, after mixing, has these color values;
e. Have an argillic or kandic horizon that has a color value, dry, of 5 or more in some subhorizon, or a color value, moist, that is less than the value, dry, by more than one unit unless the hue in some part of the argillic or kandic horizon is 5YR or yellower; and
f. Do not have a horizon within 150 cm of the surface that has >5 percent plinthite by volume.
Aquic Kandiudalfs are like Typic Kandiudalfs except for *a*, with or without *d*.
Arenic Kandiudalfs are like Typic Kandiudalfs except for *b*.
Arenic Plinthic Kandiudalfs are like Typic Kandiudalfs except for *b* and *f*.
Grossarenic Kandiudalfs are like Typic Kandiudalfs except for *c*.
Grossarenic Plinthic Kandiudalfs are like Typic Kandiudalfs except for *c* and *f*.
Mollic Kandiudalfs are like Typic Kandiudalfs except for *d*.
Plinthaquic Kandiudalfs are like Typic Kandiudalfs except for *a* and *f*.
Plinthic Kandiudalfs are like Typic Kandiudalfs except for *f*.
Rhodic Kandiudalfs are like Typic Kandiudalfs except for *d* and *e*.

Kanhapludalfs

Distinctions between Typic Kanhapludalfs and other subgroups

Typic Kanhapludalfs are the Kanhapludalfs that
a. Do not have mottles that have chroma of 2 or less within 75 cm of the soil surface if the mottled horizons are saturated with water at some time when the soil temperature at that depth is 5°C or higher or the soil has artificial drainage;
b. Do not have a lithic contact within 50 cm of the soil surface; and

c. Have an argillic or a kandic horizon that has a color value, dry, of 5 or more in some subhorizon or a color value, moist, that is less than the value, dry, by more than 1 unit, unless the hue in some part of the argillic or kandic horizon is 5YR or yellower.
Aquic Kanhapludalfs are like Typic Kanhapludalfs except for *a*.
Lithic Kanhapludalfs are like Typic Kanhapludalfs except for *b*.
Rhodic Kanhapludalfs are like Typic Kanhapludalfs except for *c*.

Natrudalfs

Distinctions between Typic Natrudalfs and other subgroups

Typic Natrudalfs are the Natrudalfs that
a. Have mottles that have chroma of 2 or less within 25 cm of the upper boundary of the natric horizon;
b. Have an Ap horizon that has a color value, moist, of 3 or more, or the soil to a depth of 18 cm, after mixing, has that color value;
c. Do not have the following combination of characteristics:
    (1) Cracks at some period in most years that are 1 cm or more wide at a depth of 50 cm, that are at least 30 cm long in some part and that extend to the surface or to the base of an Ap horizon;
    (2) A coefficient of linear extensibility (COLE) of 0.09 or more in a horizon or horizons at least 50 cm thick and a potential linear extensibility of 6 cm or more in the upper 1 m of the soil or in the whole soil if a lithic or paralithic contact is deeper than 50 cm but shallower than 1 m; and
    (3) More than 35 percent clay in horizons that total >50 cm in thickness; and
d. Do not have tonguing or interfingering of albic materials more than 2.5 cm into the natric horizon.
Glossic Natrudalfs are like Typic Natrudalfs except for *d*.
Vertic Natrudalfs are like Typic Natrudalfs except for *c*.

Paleudalfs

Distinctions between Typic Paleudalfs and other subgroups

Typic Paleudalfs are the Paleudalfs that
a. Do not have mottles that have chroma of 2 or less within 75 cm of the soil surface if the mottled horizons are saturated with water at some time when the soil temperature at that depth is 5°C or higher or the soil has artificial drainage;
b. Have texture finer than loamy fine sand in some subhorizon within 50 cm of the surface;
c. Have an Ap horizon that has a color value, moist, of 3.5 or more or a color value, dry, more than 1 unit higher than the value, moist, or the upper soil to a depth of 18 cm, after mixing, has these colors;
d. Have an argillic horizon that has a color value, dry, of 4.5 or more in some subhorizon, or a color value, moist, that is less than the value, dry, by more than one unit unless the hue in some part of the argillic horizon is 5YR or yellower;
e. Do not have the following combination of characteristics:

(1) Cracks at some period in most years that are 1 cm or more wide at a depth of 50 cm, that are at least 30 cm long in some part and that extend upward to the surface or to the base of an Ap horizon;

(2) A coefficient of linear extensibility (COLE) of 0.09 or more in a horizon or horizons at least 50 cm thick and a potential linear extensibility of 6 cm or more in the upper 1 m of the soil or in the whole soil if a lithic or paralithic contact is deeper than 50 cm but shallower than 1 m; and

(3) More than 35 percent clay in horizons that total >50 cm in thickness;

f. Have ≥ 5 percent plinthite (by volume) in all subhorizons within 1.5 m of the surface;

g. Have an argillic horizon that is continuous horizontally, that is continuous vertically for at least the upper 20 cm, and that has a texture finer than loamy fine sand;

h. Do not have subhorizons in the upper part of the argillic horizon that have skeletans that

(1) Have moist chroma of 2 or less; and

(2) Occupy 5 percent or more of the volume of the subhorizon;

i. Do not have albic materials that constitute as much as 5 percent of any subhorizon of the argillic horizon.

Albaquic Paleudalfs are like Typic Paleudalfs except for *a* and have an increase of 15 percent clay or more (absolute) within a vertical distance of 2.5 cm at the upper boundary of the argillic horizon.

Aquic Paleudalfs are like Typic Paleudalfs except for *a*, with or without *c*.

Arenic Paleudalfs are like Typic Paleudalfs except for *b* and have a sandy epipedon that is between 50 cm and 1 m thick.

Arenic Plinthic Paleudalfs are like Typic Paleudalfs except for *b* and *f* and have a sandy epipedon that is between 50 cm and 1 m thick.

Glossaquic Paleudalfs are like the Typic Paleudalfs except for *a* and *h*, with or without *c* or *i*, or both.

Glossic Paleudalfs are like Typic Paleudalfs except for *h* or *i*, or both.

Grossarenic Paleudalfs are like Typic Paleudalfs except for *b* and have a sandy epipedon that is >1 m thick.

Grossarenic Plinthic Paleudalfs are like Typic Paleudalfs except for *b* and *f* and have a sandy epipedon that is >1 m thick.

Mollic Paleudalfs are like Typic Paleudalfs except for *c*.

Plinthaquic Paleudalfs are like Typic Paleudalfs except for *a* and *f* with or without *c*.

Plinthic Paleudalfs are like Typic Paleudalfs except for *f* with or without *c*.

Psammentic Paleudalfs are like Typic Paleudalfs except for *g*, with or without *b* or *c*, or both.

Rhodic Paleudalfs are like Typic Paleudalfs except for *c* and *d*.

Vertic Paleudalfs are like Typic Paleudalfs except for *e*, with or without *a* or *c*, or both.

Rhodudalfs

Rhodudalfs are dark red Udalfs of midlatitudes that have a thinner solum than the Paleudalfs. The definition is parallel to that of other rhodic great groups. Their parent materials are basic. These soils are rare in the United States. Definitions of subgroups have not been developed.

Ustalfs

Key to great groups

HCA. Ustalfs that have a duripan that has its upper boundary within 1 m of the surface.

Durustalfs, p. 82

HCB. Other Ustalfs that have plinthite that forms a continuous phase or constitutes more than half the matrix within some subhorizon of the argillic horizon within 150 cm of the surface.

Plinthustalfs, p. 87

HCC. Other Ustalfs that have a natric horizon.

Natrustalfs, p. 85

ALF

HCD. Other Ustalfs that
1. Have a CEC <16 cmol(+) kg^{-1} clay (by 1 M NH_4OAc pH 7) and an ECEC <12 cmol(+) kg^{-1} clay (sum of bases extracted with 1 M NH_4OAc pH 7 plus 1 M KCl-extractable Al) in the major part of the argillic or kandic horizon or the major part of the upper 100 cm of the argillic or kandic horizon if these horizons are thicker than 100 cm;
2. Do not have a lithic, paralithic or petroferric contact within 150 cm of the soil surface; and
3. Have a clay distribution such that the percentage of clay does not decrease from its maximum amount by as much as 20 percent within a depth of 150 cm from the soil surface or the layer in which the clay percentage decreases by more than 20 percent has at least 5 percent of the volume consisting of skeletans on faces of peds and there is at least 3 percent (absolute) increase in clay below the layer.

Kandiustalfs, p. 83

HCE. Other Ustalfs that have a CEC <16 cmol(+) kg^{-1} clay (by 1 M NH_4OAc pH 7) and an ECEC <12 cmol(+) kg^{-1} clay (sum of bases extracted with 1 M NH_4OAc pH 7 plus 1 M KCl-extractable Al) in the major part of the argillic or kandic horizon or the major part of the upper 100 cm of the argillic or kandic horizon if these horizons are thicker than 100 cm.

Kanhaplustalfs, p. 84

HCF. Other Ustalfs that either have a petrocalcic horizon that has its upper boundary within 1.5 m of the surface or;
   1. Does not have a lithic or paralithic contact within 1.5 m of the surface; and the argillic horizon
      a. Has a clay distribution such that the percentage of clay does not decrease by as much as 20 percent of the maximum within a depth of 150 cm from the soil surface, or the layer in which the clay percentage decreases by more than 20 percent has at least 5 percent of the volume consisting of skeletans on faces of peds and there is at least 3 percent (absolute) increase in clay below the layer; and
      b. Has one or more of the following:

(1) Hues redder than 10YR and chroma of more than 4 in the matrix of at least the lower part of the horizon;
(2) Hues of 7.5YR or redder and value, moist, that is less than 4 and value, dry, that is less than 5 throughout the major part of the horizon; or
(3) Common coarse mottles that have hue of 7.5YR or redder or chroma of more than 5 in the lower part of the horizon; or
2. Does not have a lithic or paralithic contact within 50 cm of the surface and has an argillic horizon in which the upper part has a clayey particle-size class and there is an increase of at least 20 percent clay (absolute) within a vertical distance of 7.5 cm, or of at least 15 percent clay (absolute) within a vertical distance of 2.5 cm at the upper boundary.

Paleustalfs, p. 86

HCG. Other Ustalfs that have an argillic horizon that has throughout its thickness a hue redder than 5YR, a color value, moist, less than 3.5, and a color value, dry, no more than one unit higher than the value moist.

Rhodustalfs, p. 88

HCH. Other Ustalfs.

Haplustalfs, p. 82

Durustalfs

Durustalfs are the Ustalfs that have a duripan whose upper boundary is within 1 m of the surface. They are not known to occur in the United States, and subgroups have not been developed. They are provided for use in other countries.

Haplustalfs

Distinctions between Typic Haplustalfs and other subgroups

Typic Haplustalfs are the Haplustalfs that
a. Do not have mottles that have chroma of 2 or less within 75 cm of the soil surface if the mottled horizon is saturated with water at some time during the year or the soil has artificial drainage;
b. Have texture finer than loamy fine sand in some subhorizon within 50 cm of the soil surface;
c. Do not have a lithic contact within 50 cm of the soil surface;
d. Have CEC of 24 or more cmol(+) kg^{-1} clay (by 1 M NH$_4$OAc pH7);
e. Have an argillic horizon that is continuous horizontally, that is continuous vertically for at least the upper 20 cm, that is not composed entirely of lamellae, and that has a texture finer than loamy fine sand;
f. When neither irrigated nor fallowed to store moisture:
(1) If the soil temperature regime is mesic or thermic, are dry for more than four-tenths of the cumulative days in some part of the moisture control section when the soil temperature at a depth of 50 cm exceeds 5°C; or
(2) If the soil temperature regime is hyperthermic, isomesic, or warmer, the soils are dry in some or all parts of the moisture control section for more than 90 days

during a period when the soil temperature at a depth of 50 cm exceeds 8°C;

g. Have an argillic horizon that has base saturation (by sum of cations) of 75 percent or more in some part;

h. Do not have the following combination of characteristics:

(1) Cracks at some period in most years that are 1 cm or more wide at a depth of 50 cm, that are at least 30 cm long in some part, and that extend to the surface or to the base of an Ap horizon if the soil is not irrigated;

(2) A coefficient of linear extensibility (COLE) of 0.07 or more in a horizon or horizons at least 50 cm thick and a potential linear extensibility of 6 cm or more in the upper 1.25 m of the soil or in the whole soil if a lithic or paralithic contact is deeper than 50 cm but shallower than 1.25 m; and

(3) More than 35 percent clay in horizons that total >50 cm in thickness; and

i. When neither irrigated nor fallowed to store moisture:

(1) If the soil temperature regime is mesic or thermic, are dry less than six-tenths of the time in half or more years in some part of the moisture control section (not necessarily the same part) during a period when the soil temperature at a depth of 50 cm exceeds 5°C; or

(2) If the soil temperature regime is hyperthermic, or isomesic or a warmer iso-temperature regime, the soils are moist in most years in some or all parts of the mois-ture control section for 90 consecutive days or more during a period when the soil temperature at a depth of 50 cm exceeds 8°C.

Aquic Haplustalfs are like Typic Haplustalfs except for *a* or for *a* and *f*.

Aquic Arenic Haplustalfs are like Typic Haplustalfs except for *a* and *b*, with or without *f* or *g*, or both.

Aquultic Haplustalfs are like Typic Haplustalfs except for *a* and *g*, with or without *f*.

Arenic Haplustalfs are like Typic Haplustalfs except for *b*, with or without *f* or *g*, or both.

Arenic Aridic Haplustalfs are like Typic Haplustalfs except for *b* and *i*.

Aridic Haplustalfs are like Typic Haplustalfs except for *i*.

Kanhaplic Haplustalfs are like Typic Haplustalfs except for *d*, with or without *g* or *f*, or both.

Lithic Haplustalfs are like Typic Haplustalfs except for *c*, with or without *f* or *i*, or both.

Psammentic Haplustalfs are like Typic Haplustalfs except for *e*, with or without any or all of *b*, *f*, and *g*.

Udertic Haplustalfs are like Typic Haplustalfs except for *f* and *h*, with or without *a*.

Udic Haplustalfs are like Typic Haplustalfs except for *f*.

Ultic Haplustalfs are like Typic Haplustalfs except for *g* or for *f* and *g*.

Vertic Haplustalfs are like Typic Haplustalfs except for *h*, with or without *a* or *i* or both.

Kandiustalfs

Distinctions between Typic Kandiustalfs and other sub-groups

Typic Kandiustalfs are the Kandiustalfs that

a. Do not have mottles that have chroma of 2 or less within 75 cm of the soil surface if the mottled horizon is saturated with water at some time of the year when the soil temperature at that depth is 5°C or higher or the soil has artificial drainage;
b. Do not have an epipedon that is 50 cm to 100 cm thick if the particle-size class is sandy throughout;
c. Do not have an epipedon that is >100 cm thick if the particle-size class is sandy throughout;
d. Do not have a horizon within 150 cm of the soil surface that has >5 percent plinthite by volume;
e. When neither irrigated nor fallowed to store moisture:
    (1) If the soil temperature regime is mesic or thermic, are moist more than six-tenths of the time in half or more years in some part of the moisture control section (not necessarily the same part) when the soil temperature at a depth of 50 cm exceeds 5°C; or
    (2) If the soil temperature regime is hyperthermic, isomesic, or warmer, are moist in most years in some or all parts of the moisture control section for 180 or more days during a period when the soil temperature at a depth of 50 cm exceeds 8°C; and
f. When neither irrigated nor fallowed to store moisture:
    (1) If the soil temperature regime is mesic or thermic, are dry for more than 135 cumulative days in some part of the moisture control section when the soil temperature at a depth of 50 cm exceeds 5°C; or
    (2) If the soil temperature regime is hyperthermic, isomesic, or warmer, the soils are dry in some or all parts of the moisture control section for more than 90 days during a period when the soil temperature at a depth of 50 cm exceeds 8°C; and
g. Have an argillic or kandic horizon that has a color hue of 5YR or yellower in some part, or has a value, moist, of 3.5 or more in some part, or has a value, dry, that is more than one unit higher than the value, moist.

Aquic Kandiustalfs are like Typic Kandiustalfs except for *a*, with or without *e*.
Aquic Arenic Kandiustalfs are like Typic Kandiustalfs except for *a* and *b*.
Arenic Kandiustalfs are like Typic Kandiustalfs except for *b*.
Arenic Aridic Kandiustalfs are like Typic Kandiustalfs except for *b* and *e*.
Aridic Kandiustalfs are like Typic Kandiustalfs except for *e*.
Grossarenic Kandiustalfs are like *Typic Kandiustalfs* except for *c*, with or without *a*.
Rhodic Kandiustalfs are like Typic Kandiustalfs except for *g*.
Udic Kandiustalfs are like Typic Kandiustalfs except for *f*.

Kanhaplustalfs

Distinctions between Typic Kanhaplustalfs and other subgroups

Typic Kanhaplustalfs are the Kanhaplustalfs that
a. Do not have mottles that have chroma of 2 or less within 75 cm of the soil surface if the mottled horizon is saturated

with water at sometime when the soil temperature at that
depth is 5°C or higher or the soil has artificial drainage;
b. Do not have a lithic contact within 50 cm of the soil
surface;
c. When neither irrigated nor fallowed to store moisture:
    (1) If the soil temperature regime is mesic or thermic,
    are moist more than six-tenths of the time in half or
    more years in some part of the moisture control section
    (not necessarily the same part) when the soil temperature
    at a depth of 50 cm exceeds 5°C; or
    (2) If the soil temperature regime is hyperthermic,
    isomesic, or warmer, are moist in most years in some or
    all parts of the moisture control section for 180 or more
    days during a period when the soil temperature at a
    depth of 50 cm exceeds 8°C; and
d. When neither irrigated nor fallowed to store moisture:
    (1) If the soil temperature regime is mesic or thermic,
    are dry for more than 135 cumulative days in some part
    of the Moisture control section when the soil temperature
    at a depth of 50 cm exceeds 5°C; or
    (2) If the soil temperature regime is hyperthermic,
    isomesic, or warmer, the soils are dry in some or all parts
    of the moisture control section for more than 90 days
    during a period when the soil temperature at a depth of
    50 cm exceeds 8°C; and
e. Have an argillic or kandic horizon that has a color hue of
5YR or yellower in some part, or has a value, moist, of 3.5
or more in some part, or has a value, dry, that is more than
one unit higher than the value, moist.
Aquic Kanhaplustalfs are like Typic Kanhaplustalfs except
for *a.*
Aridic Kanhaplustalfs are like Typic Kanhaplustalfs except
for *c.*
Lithic Kanhaplustalfs are like Typic Kanhaplustalfs except
for *b* with or without *c.*
Rhodic Kanhaplustalfs are like Typic Kanhaplustalfs except
for *e.*
Udic Kanhaplustalfs are like Typic Kanhaplustalfs except
for *d.*

Natrustalfs

Distinctions between Typic Natrustalfs and other subgroups

Typic Natrustalfs are the Natrustalfs that
a. Do not have mottles that have chroma of 2 or less within
50 cm of the soil surface if there is ground water in the
mottled horizon at some time of year when the soil temperature is 5°C or higher;
b. Have an Ap horizon that has a color value, moist, more
than 3, or the surface soil to a depth of 18 cm, after mixing,
has a color value, moist, more than 3;
c. Do not have a salic horizon that has its upper boundary
within 75 cm of the soil surface; and
d. Do not have a petrocalcic horizon that has its upper
boundary within 1.5 m of the surface.
Aquic Natrustalfs are like Typic Natrustalfs except for *a,*
with or without *b.*
Mollic Natrustalfs are like Typic Natrustalfs except for *b.*
Petrocalcic Natrustalfs are like Typic Natrustalfs except for
d.

Salorthidic Natrustalfs are like Typic Natrustalfs except for
c.

Paleustalfs

Distinctions between Typic Paleustalfs and other subgroups

Typic Paleustalfs are the Paleustalfs that
a. Do not have mottles that have chroma of 2 or less within
75 cm of the soil surface if the mottled horizon is saturated
with water at some time of the year when the temperature
of the horizon is 5°C or higher;
b. Have texture finer than loamy fine sand in some subhorizon within 50 cm of the soil surface;
c. Have <5 percent plinthite by volume in all subhorizons
within 1.5 m of the soil surface;
d. When neither irrigated nor fallowed to store moisture:
   (1) If the soil temperature regime is mesic or thermic,
   are dry for more than four-tenths of the cumulative days
   in some part of the moisture control section when the soil
   temperature at a depth of 50 cm exceeds 5°C; or
   (2) If the soil temperature regime is hyperthermic,
   isomesic, or warmer, the soils are dry in some or all parts
   of the moisture control section for more than 90 days
   during a period when the soil temperature at a depth of
   50 cm exceeds 8°C;
e. Have an argillic horizon that has base saturation (by sum
of cations) of 75 percent or more in some part;
f. Do not have the following combination of characteristics:
   (1) Cracks at some period in most years that are 1 cm or
   more wide at a depth of 50 cm, that are at least 30 cm
   long in some part, and that extend upward to the surface
   or to the base of an Ap horizon if the soil is not irrigated;
   (2) A coefficient of linear extensibility (COLE) of 0.07
   or more in a horizon or horizons at least 50 cm thick and
   a potential linear extensibility of 6 cm or more in the
   upper 1.25 m of the soil or in the whole soil if a lithic or
   paralithic contact is deeper than 50 cm but shallower than
   1.25 m; and
   (3) More than 35 percent clay in horizons that total >50
   cm in thickness;
g. Do not have a petrocalcic horizon that has its upper
boundary within 1.5 m of the soil surface;
h. When not irrigated and when not fallowed to store moisture:
   (1) If the soil temperature regime is mesic or thermic,
   are dry less than six-tenths of the time in half or more
   years in some part of the moisture control section (not
   necessarily the same part) during a period when the soil
   temperature at a depth of 50 cm exceeds 5°C; or
   (2) If the soil temperature regime is hyperthermic,
   isomesic, or warmer, are moist in most years in some or
   all parts of the moisture control section for 90 consecutive days or more during a period when the soil temperature at a depth of 50 cm exceeds 8°C;
i. Have CEC of 24 or more cmol(+) kg^{-1} clay (by 1 M
NH$_4$OAc pH7) in the major part of the argillic horizon or
the major part of the upper 100 cm of the argillic horizon is
thicker than 100 cm;

j. Have an argillic horizon that has a color hue of 5YR or yellower in some part, or has a value, moist, of 3.5 or more in some part, or has a value, dry, that is more than one unit higher than the value, moist; and

k. Have an argillic horizon that is continuous horizontally, that is continuous vertically for at least the upper 20 cm, and that has a texture finer than loamy fine sand.

Aquic Paleustalfs are like Typic Paleustalfs except for *a*, with or without *d* or *e*, or both.

Aquic Arenic Paleustalfs are like Typic Paleustalfs except for *a* and *b*, with or without *d* or *e*, or both, and they have a sandy epipedon (loamy fine sand or coarser) between 50 cm and 1 m thick.

Arenic Paleustalfs are like Typic Paleustalfs except for *b* or for *b* and *d*, and they have a sandy epipedon (loamy fine sand or coarser) between 50 cm and 1 m thick.

Arenic Aridic Paleustalfs are like Typic Paleustalfs except for *b* and *h* and they have a sandy epipedon (loamy fine sand or coarser) between 50 cm and 1 m thick.

Aridic Paleustalfs are like Typic Paleustalfs except for *h*, and either they are noncalcareous in some subhorizon above the calcic horizon or they do not have a calcic horizon whose upper boundary is within a depth of 1 m below the soil surface if the weighted average particle-size class of the upper 50 cm of the argillic horizon is sandy, or 60 cm if it is loamy, or 50 cm if it is clayey.

Calciorthidic Paleustalfs are like Typic Paleustalfs except for *h*, and they have a calcic horizon within a depth of 1 m if the weighted average particle-size class of the upper 50 cm of the argillic horizon is sandy, 60 cm if it is loamy, or 50 cm if it is clayey, and they have carbonates in all sub-horizons above the calcic horizon.

Grossarenic Paleustalfs are like Typic Paleustalfs except for *b*, with or without *a* or *d*, or both, and they have a sandy epipedon (loamy fine sand or coarser) >1 m thick.

Kandic Paleustalfs are like Typic Paleustalfs except for *i*, with or without *d* or *e*, or both.

Petrocalcic Paleustalfs are like Typic Paleustalfs except for *g*, with or without *h* or *j*, or both.

Psammentic Paleustalfs are like Typic Paleustalfs except for *k*, with or without some or all of *b*, *d*, or *e*.

Rhodic Paleustalfs are like Typic Paleustalfs except for *j*, with or without *d* or *e* or both, and the hue in the argillic horizon is redder than 5YR in all parts.

Udertic Paleustalfs are like Typic Paleustalfs except for *d* and *f*, with or without *a*.

Udic Paleustalfs are like Typic Paleustalfs except for *d*.

Ultic Paleustalfs are like Typic Paleustalfs except for *e* and *d*.

Plinthustalfs

Plinthustalfs are the Ustalfs that have plinthite that forms a continuous phase or that constitutes more than half the matrix of some subhorizon of the argillic horizon within 1.25 m of the soil surface. There are no soil series in the United States that are presently classified in this great group, but the group is provided for other parts of the world. Subgroups have not been developed. Plinthustalfs were included with Ground-Water Laterite soils in the 1938 classification.

Rhodustalfs

Distinctions between Typic Rhodustalfs and other subgroups

Because these soils are rare in the United States, the classification that follows probably is incomplete, and it is provisional.

Typic Rhodustalfs are the Rhodustalfs that
a. Do not have a lithic contact within 50 cm of the soil surface;
b. Have CEC of 24 or more cmol(+) kg^{-1} (by 1 M NH$_4$OAc pH7) in the major part of the argillic horizon or the major part of the upper 100 cm of the argillic horizon if the argillic horizon is thicker than 100 cm; and
c. When neither irrigated nor fallowed to store moisture:
   (1) If the soil temperature regime is mesic or thermic, are dry for more than four-tenths of the cumulative days in some part of the moisture control section when the soil temperature at a depth of 50 cm exceeds 5°C; or
   (2) If the soil temperature regime is hyperthermic, isomesic, or warmer, the soils are dry in some or all parts of the moisture control section for more than 90 days during a period when the soil temperature at a depth of 50 cm exceeds 8°C;
Kandic Rhodustalfs are like Typic Rhodustalfs except for *b*, with or without *c*.
Lithic Rhodustalfs are like Typic Rhodustalfs except for *a*, with or without *c*.
Udic Rhodustalfs are like Typic Rhodustalfs except for *c*.

Xeralfs

Key to great groups

HDA. Xeralfs that have a duripan whose upper boundary is within 1 m of the soil surface but below an argillic or a natric horizon.

Durixeralfs, p. 89

HDB. Other Xeralfs that have a natric horizon.

Natrixeralfs, p. 91

HDC. Other Xeralfs that have a fragipan.

Fragixeralfs, p. 89

HDD. Other Xeralfs that have plinthite that forms a continuous phase in or constitutes more than half the matrix of some subhorizon of the argillic horizon within 150 cm of the soil surface.

Plinthoxeralfs, p. 92

HDE. Other Xeralfs that have an argillic horizon that, in all parts, has a color hue redder than 5YR and a value, moist, less than 3.5 and a value, dry, no more than one unit higher than the value, moist.

Rhodoxeralfs, p. 92

HDF. Other Xeralfs that have either a petrocalcic horizon whose upper boundary is within 1.5 m of the soil surface or have one or both of

1. A vertical clay distribution such that the percentage of clay does not decrease from the maximum by as much as 20 percent of that maximum throughout a depth of 1.5 m from the soil surface, or the horizon in which the clay decreases has either

   a. More than 5 percent by volume of plinthite; or
   b. Skeletans or other evidences of clay eluviation, and there is no lithic or paralithic contact within 1.5 m of the soil surface, and either
   (1) A hue redder than 10YR and chroma more than 4 in the matrix of at least the lower part of the argillic horizon; or
   (2) Common coarse mottles that have a hue of 7.5YR or redder or chroma greater than 5 or both in the lower part of the argillic horizon;

2. An argillic horizon that has a clayey particle-size class in the upper part and an increase of at least 20 percent clay (absolute) within a vertical distance of 7.5 cm or at least 15 percent clay (absolute) within a vertical distance of 2.5 cm at the upper boundary; and there is no lithic or paralithic contact within 50 cm of the surface of the soil.

Palexeralfs, p. 91

HDG. Other Xeralfs.

Haploxeralfs, p. 90

Durixeralfs

Distinctions between Typic Durixeralfs and other subgroups

Typic Durixeralfs are the Durixeralfs that
a. Have an argillic horizon that has <35 percent clay in all parts, or the increase in clay content is <15 percent clay (absolute) within a vertical distance of 2.5 cm at the upper boundary of the argillic horizon, or the increase is <10 percent clay (absolute) if the soil is cultivated and the lower boundary of the Ap horizon is the upper boundary of the argillic horizon;
b. Do not have mottles in the argillic horizon that have chroma of 2 or less;
c. Have a duripan that is massive, platy, or prismatic and that has half or more of its upper boundary indurated and coated with opal or with opal and sesquioxides or that is indurated in some subhorizon below the upper boundary; and
d. Do not have a natric horizon.
Abruptic Durixeralfs are like Typic Durixeralfs except for *a.*
Abruptic Haplic Durixeralfs are like Typic Durixeralfs except for *a* and *c.*
Haplic Durixeralfs are like Typic Durixeralfs except for *c.*
Natric Durixeralfs are like Typic Durixeralfs except for *d,* with or without *b* or *c,* or both.

Fragixeralfs

Distinctions between Typic Fragixeralfs and other subgroups

Typic Fragixeralfs are the Fragixeralfs that
a. Above the fragipan have an argillic horizon that has clay skins on at least some vertical and horizontal faces of primary or secondary peds, or both;

b. Do not have a layer in the upper 75 cm that has a texture finer than loamy fine sand, that is as much as 18 cm thick, that has a bulk density (at 33 kPa water tension) of 0.95 g per cubic centimeter or less in the fine-earth fraction, and that has either

   (1) a ratio of measured clay to 1500 kPa water (percentages) of 1.25 or less or

   (2) a ratio of CEC (at pH near 8) to 1500 kPa water of 1.5 and more exchange acidity than the sum of bases plus KCl-extractable aluminum;

c. Either have an Ap horizon that has a color value, moist, of 4 or more or a color value, dry, of 6 or more, when crushed and smoothed, or the soil to a depth of 18 cm, after mixing, has those colors; and

d. Do not have mottles that have chroma of 2 or less in the upper 25 cm of the argillic horizon and do not have mottles that have chroma of 2 or less within 40 cm of the surface if the horizons that have mottles of low chroma are saturated with water at some time of the year when the soil temperature is 5°C or higher in those horizons. Mottles are not the same as skeletans that may also have low chroma.

Mollic Fragixeralfs are like Typic Fragixeralfs except for *c*, with or without *d*.

Ochreptic Fragixeralfs are like Typic Fragixeralfs except for *a*.

Haploxeralfs

Distinctions between Typic Haploxeralfs and other subgroups

Typic Haploxeralfs are the Haploxeralfs that

a. Do not have mottles that have chroma of 2 or less within 75 cm of the soil surface if the mottled horizon is saturated with water at some time when the temperature of that horizon is 5°C or higher or there is artificial drainage;

b. Have an A horizon that throughout its upper 10 cm has a color value, moist, of 3.5 or more or has <0.7 percent organic carbon in some part, or have an Ap horizon that has a color value, moist, of 3.5 or more or contains <0.7 percent organic carbon;

c. Do not have a lithic contact within 50 cm of the soil surface;

d. Have exchangeable sodium that is <15 percent of the CEC (at pH 8.2) throughout the argillic horizon;

e. Have <5 percent plinthite (by volume) in all subhorizons within 1.5 m of the soil surface;

f. Have an argillic horizon that has base saturation (by sum of cations) of 75 percent or more throughout the upper 75 cm or to a lithic or paralithic contact, whichever is shallower;

g. Do not have the following combination of characteristics:

   (1) Cracks at some period in most years that are 1 cm or more wide at a depth of 50 cm, that are at least 30 cm long in some part, and that extend upward to the surface or to the base of an Ap horizon if the soil is not irrigated;

   (2) A coefficient of linear extensibility (COLE) of 0.05 or more in a horizon or horizons at least 50 cm thick and a potential linear extensibility of 6 cm or more in the upper 1.5 m of the soil or in the whole soil if a lithic or paralithic contact is deeper than 50 cm but shallower than 1.5 m; and

(3) More than 35 percent clay in horizons that total >50 cm in thickness;

h. Have an argillic horizon that is continuous vertically for at least the upper 20 cm that is not composed entirely of lamellae, and that has a texture finer than loamy fine sand;

i. Do not have a calcic horizon that has its upper boundary within the upper 1 m of soil; and

j. Have an argillic horizon that is continuous horizontally throughout the area of each pedon.

Aquic Haploxeralfs are like Typic Haploxeralfs except for *a* or for *a* and *b*.

Aquultic Haploxeralfs are like Typic Haploxeralfs except for *a* and *f* with or without *b*.

Calcic Haploxeralfs are like Typic Haploxeralfs except for *i*.

Lithic Haploxeralfs are like Typic Haploxeralfs except for *c*.

Lithic Mollic Haploxeralfs are like Typic Haploxeralfs except for *c* and *b*.

Lithic Ruptic-Xerochreptic Haploxeralfs are like Typic Haploxeralfs except for *c* and *j*.

Mollic Haploxeralfs are like Typic Haploxeralfs except for *b*.

Natric Haploxeralfs are like Typic Haploxeralfs except for *d*.

Psammentic Haploxeralfs are like Typic Haploxeralfs except for *h*, with or without *b* or *f*, or both.

Ultic Haploxeralfs are like Typic Haploxeralfs except for *f* or for *f* and *b*.

Vertic Haploxeralfs are like Typic Haploxeralfs except for *g* or for *b* and *g*.

Natrixeralfs

Distinctions between Typic Natrixeralfs and other subgroups

Typic Natrixeralfs are the Natrixeralfs that

a. Do not have mottles that have chroma of 2 or less within 75 cm of the soil surface if there is ground water in the mottled horizon at some time when the temperature of that horizon is 5°C or higher.

Aquic Natrixeralfs are like Typic Natrixeralfs except for *a*.

Palexeralfs

Distinctions between Typic Palexeralfs and other subgroups

The list of subgroups is incomplete for the world. A few subgroups are defined that are not known to occur in the United States, but a number of others that have not yet been defined are known to exist.

Typic Palexeralfs are the Palexeralfs that

a. Do not have mottles that have chroma of 2 or less within 75 cm of the soil surface if the mottled horizon is saturated with water at some time of the year when its temperature is 5°C or higher or there is artificial drainage;

b. Do not have a calcic horizon within 1.5 m of the soil surface;

c. Have an A horizon that, throughout its upper 10 cm, has a color value, moist, of 3.5 or more or contains <0.7 percent organic carbon, or they have an Ap horizon that has a color value, moist, of 3.5 or more or contains <0.7 percent organic carbon;

d. Do not have a petrocalcic horizon whose upper boundary is within 1.5 m of the soil surface;

e. Have an argillic horizon that has at least 75 percent base saturation (by sum of cations) in some part;

f. Have an argillic horizon in which the upper part has a clayey particle-size class and there is an increase of at least 20 percent clay (absolute) within a vertical distance of 7.5 cm or of at least 15 percent clay (absolute) within 2.5 cm at the upper boundary;

g. Have <5 percent plinthite (by volume) in all subhorizons within 1.5 m of the soil surface;

h. Do not have the following combination of characteristics:
   (1) Cracks at some period in most years that are 1 cm or more wide at a depth of 50 cm, that are at least 30 cm long in some part and that extend upward to the surface or to the base of an Ap horizon if the soil is not irrigated; and
   (2) A coefficient of linear extensibility (COLE) of 0.05 or more in a horizon or horizons at least 50 cm thick and a potential linear extensibility of 6 cm or more in the upper 1.5 m of the soil or in the whole soil if a lithic or paralithic contact is deeper than 50 cm but shallower than 1.5 m, and
   (3) More than 35 percent clay in horizons that total >50 cm in thickness;

i. Have texture finer than loamy fine sand in some subhorizon within 50 cm of the soil surface; and

j. Have <15 percent saturation with sodium in all subhorizons within 1 m of the soil surface.

Aquic Palexeralfs are like Typic Palexeralfs except for *a* or for *a* and *c*.

Calcic Palexeralfs are like Typic Palexeralfs except for *b*.

Mollic Palexeralfs are like Typic Palexeralfs except for *c*.

Natric Palexeralfs are like Typic Palexeralfs except for *j*.

Petrocalcic Palexeralfs are like Typic Palexeralfs except for *d*, or for *d* and *f*.

Ultic Palexeralfs are like Typic Palexeralfs except for *e*, with or without *f* or *c*, or both.

Vertic Palexeralfs are like Typic Palexeralfs except for *h*, with or without *a* or *c*, or both.

Plinthoxeralfs

Plinthoxeralfs are the Xeralfs that have plinthite that forms a continuous phase or that constitutes more than half the matrix of some subhorizon within 150 cm of the soil surface. Few of these soils are in the United States, but the soils are moderately extensive in some parts of the world. Subgroups have not been developed. The Plinthoxeralfs were included with Ground-Water Laterite soils in the 1938 classification.

Rhodoxeralfs

Distinctions between Typic Rhodoxeralfs and other subgroups

The list of subgroups that follows is incomplete because the
soils are of such limited extent in the United States.

Typic Rhodoxeralfs are the Rhodoxeralfs that
a. Have an argillic horizon that is >15 cm thick and is continuous in each pedon;
b. Do not have a lithic contact within 50 cm of the soil
surface;
c. Do not have a petrocalcic horizon whose upper boundary
is within 1.5 m of the soil surface; and
d. Do not have a calcic horizon whose upper boundary is
within 1.5 m of the soil surface.
Calcic Rhodoxeralfs are like Typic Rhodoxeralfs except for
d.
Lithic Rhodoxeralfs are like Typic Rhodoxeralfs except for
b.
Petrocalcic Rhodoxeralfs are like Typic Rhodoxeralfs except
for *c.*

Chapter 6
Aridisols

Key To Suborders

EA. Aridisols that have an argillic or a natric horizon.

Argids, p. 95

EB. Other Aridisols.

Orthids, p. 102

Argids

Key to great groups

EAA. Argids that have a duripan[1] below an argillic horizon and do not have a natric horizon.

Durargids, p. 95

EAB. Other Argids that have a duripan below a natric horizon.

Nadurargids, p. 99

EAC. Other Argids that have a natric horizon and do not have a petrocalcic horizon.

Natrargids, p. 99

EAD. Other Argids that do not have a lithic or paralithic contact within 50 cm of the soil surface, that have a petrocalcic horizon or that have an argillic horizon that has 35 percent or more clay in some part, and that have either:
    1. An increase of 15 percent or more clay (absolute) within a vertical distance of 2.5 cm at the upper boundary of the argillic horizon; or
    2. An increase of 10 percent or more clay (absolute) if cultivated and the lower boundary of the Ap horizon is the upper boundary of the argillic horizon.

Paleargids, p. 101

EAE. Other Argids.

Haplargids, p. 96

Durargids

Distinctions between Typic Durargids and other subgroups

Typic Durargids are the Durargids that
a. Are not saturated with water for 90 consecutive days or more within 1 m of the surface in most years and do not have any of the following characteristics within 1 m of the soil surface if there is ground water within this depth at some time in most years:
    (1) Dominant chroma of 1 or less throughout the horizons and hue as yellow or yellower than 2.5Y in some part;
    (2) Dominant chroma of 2 or less and mottles that are not due to segregated lime;

---

[1]A duripan or a petrocalcic horizon must have its upper boundary within 1 m of the surface to be diagnostic in Aridisols.

(3) Both a dominant chroma of 2 or less and a greater SAR (or percentage of exchangeable sodium) in more than half the thickness of the horizon between the surface and 50 cm depth than in the saturated zone;

b. Have a platy or massive duripan that is indurated in some subhorizon;

c. Have a duripan at a depth <18 cm; or the weighted average percentage of organic carbon in the upper soil to a depth of 40 cm is <0.6 if the weighted average ratio of sand to clay in the upper soil to that depth is 1.0 or less, or is less than one-seventh percent if the ratio is 13 or more, or is intermediate between 0.6 percent and one-seventh percent if the ratio of sand to clay in the upper soil is >1.0 but <13; or the weighted average percentage of organic carbon in the upper soil to a depth of 18 cm is not as much as one-fifth more than the values just stated if a duripan is present at a depth <40 cm but >18 cm;

d. Do not have an argillic horizon that has 35 percent or more clay in some part and also has either

(1) An increase of 15 percent or more clay (absolute) within a vertical distance of 2.5 cm at the upper boundary of the argillic horizon; or

(2) An increase of 10 percent or more clay (absolute) if cultivated and the lower boundary of the Ap horizon is the upper boundary of the argillic horizon; and

e. Are dry in all parts of the moisture control section more than three-fourths of the time (cumulative) that the soil temperature at a depth of 50 cm is 5°C or higher.

Abruptic Durargids are like Typic Durargids except for *d*.

Abruptic Xerollic Durargids are like Typic Durargids except for *c*, *d*, and *e*. They have a mean annual soil temperature lower than 22°C, the mean summer and mean winter soil temperatures at a depth of 50 cm differ by 5°C or more, and they have an aridic moisture regime that borders on a xeric regime.

Haplic Durargids are like Typic Durargids except for *b*.

Haploxerollic Durargids are like Typic Durargids except for *b* and *c*, with or without *e*. They have a mean annual soil temperature lower than 22°C, the mean summer and mean winter soil temperatures at a depth of 50 cm differ by 5°C or more, and they have an aridic moisture regime that borders on a xeric regime.

Xerollic Durargids are like Typic Durargids except for *c*, with or without *e*. They have a mean annual soil temperature lower than 22°C, the mean summer and mean winter soil temperatures at a depth of 50 cm differ by 5°C or more, and they have an aridic moisture regime that borders on a xeric regime.

Haplargids

Distinctions between Typic Haplargids and other subgroups

Typic Haplargids are the Haplargids that

a. Are not saturated with water for 90 consecutive days or more within 1 m of the surface in most years and do not have any of the following characteristics within a depth of 1 m below the surface if there is ground water within this depth at some time in most years:

(1) Dominant chroma of 1 or less throughout and a hue as yellow or yellower than 2.5Y in some part;

(2) Dominant chroma of 2 or less and mottles that are not due to segregated lime; or

(3) Both a dominant chroma of 2 or less and a greater SAR (or percentage of exchangeable Na) in more than half the thickness of the horizons between the surface and 50 cm depth than in the saturated zone;

b. Have texture finer than loamy fine sand in some subhorizon above a depth of 50 cm;

c. Do not have a horizon within 1 m of the surface that is >15 cm thick and that either contains 20 percent or more (by volume) durinodes in a nonbrittle matrix or is brittle and has firm consistence when moist;

d. Do not have a lithic contact within 50 cm of the surface;

e. Have a weighted average percentage of organic carbon in the upper 40 cm that is <0.6 percent if the weighted average ratio of sand to clay in the soil above that depth is 1.0 or less, or is less than one-seventh percent if the ratio is 13 or more, or have an intermediate percentage of organic carbon if the ratio of sand to clay is between 1.0 and 13; or have a weighted average percentage of organic carbon in the soil to a depth of 18 cm that is not as much as one-fifth more than the values just stated if there is a lithic or paralithic contact at a depth <40 cm but >18 cm;

f. Have an argillic horizon that is continuous throughout the area of each pedon;

g. Do not have the following combination of characteristics:

    (1) Cracks at some period in most years that are 1 cm or more wide at a depth of 50 cm, that are at least 30 cm long in some part, and that extend upward to the surface, to the base of an Ap horizon, or to the top of the argillic horizon;

    (2) A coefficient of linear extensibility (COLE) of 0.05 or more in a horizon or horizons at least 50 cm thick and a potential linear extensibility of 6 cm or more in the upper 1.5 m of the soil or in the whole soil if a lithic or paralithic contact is deeper than 50 cm but shallower than 1.5 m; and

    (3) More than 35 percent clay in horizons that total >50 cm in thickness; and

h. Are dry in all parts of the moisture control section more than three-fourths of the time (cumulative) that the soil temperature is 5°C or higher at a depth of 50 cm.

Aquic Haplargids are like Typic Haplargids except for *a*, with or without *e* or *h*, or both.

Arenic Haplargids are like Typic Haplargids except for *b*.

Arenic Ustalfic Haplargids are like Typic Haplargids except for *b* and *h*. They have a mean annual soil temperature of 8°C or higher and an aridic moisture regime that borders on an ustic regime.

Arenic Ustollic Haplargids are like Typic Haplargids except for *b* and *e*, with or without *h*. They have a mean annual soil temperature of 8°C or higher and an aridic moisture regime that borders on an ustic regime.

Borollic Haplargids are like Typic Haplargids except for *e*, with or without *h*. The mean annual soil temperature is lower than 8°C, and the mean summer and mean winter soil temperatures at a depth of 50 cm differ by 5°C or more. Their aridic moisture regime borders on an ustic regime.

Borollic Lithic Haplargids are like Typic Haplargids except for *d* and *e*, with or without *h*. They have a frigid temperature regime and an aridic moisture regime that borders on an ustic regime.

Borollic Vertic Haplargids are like Typic Haplargids except for *e* and *g*, with or without *h*. The mean annual soil temperature is lower than 8°C, and the mean summer and mean winter soil temperatures at a depth of 50 cm differ by 5°C or more. The cracks are not closed for as many as 60 consecutive days of the 120 days following the winter solstice in 3 or more years out of 10.

Duric Haplargids are like Typic Haplargids except for *c*.

Durixerollic Haplargids are like Typic Haplargids except for *c* and *e*, with or without *h*. They have a mean annual soil temperature lower than 22°C, and the mean summer and mean winter soil temperatures at a depth of 50 cm differ by 5°C or more. The aridic moisture regime borders on a xeric regime.

Lithic Haplargids are like Typic Haplargids except for *d*.

Lithic Ruptic-Entic Xerollic Haplargids are like Typic Haplargids except for *d*, *e*, and *f*, with or without h. They have a mean annual soil temperature lower than 22°C, and the mean summer and mean winter soil temperatures at a depth of 50 cm differ by 5°C or more. The aridic moisture regime borders on a xeric regime.

Lithic Ustollic Haplargids are like Typic Haplargids except for *d* and *e*, with or without *h*. They have a mean annual soil temperature of 8°C or higher and an aridic moisture regime that borders on an ustic regime.

Lithic Xerollic Haplargids are like Typic Haplargids except for *d* and *e*, with or without *h*. They have a mean annual soil temperature that is lower than 22°C, and the mean summer and mean winter soil temperatures at a depth of 50 cm differ by 5°C or more. They have an aridic moisture regime that borders on a xeric regime.

Ustalfic Haplargids are like Typic Haplargids except for *h*. They have a mean annual soil temperature of 8°C or higher and an aridic moisture regime that borders on an ustic regime.

Ustertic Haplargids are like Typic Haplargids except for *g*, with or without *e* or *h*, or both. They have cracks that remain open from 175 to 240 days, cumulative, in most years and that are not closed for as many as 60 consecutive days during the 120 days following the winter solstice in 3 or more years out of 10 if the soil temperature regime is thermic, mesic, or frigid. Other frigid soils are excluded.

Ustollic Haplargids are like Typic Haplargids except for *e*, with or without *h*. They have a mean annual soil temperature of 8°C or higher and an aridic moisture regime that borders on an ustic regime.

Vertic Haplargids are like Typic Haplargids except for *g* or for *g* and *h* and have cracks that remain open for 8 months or more, cumulative, in most years.

Xeralfic Haplargids are like Typic Haplargids except for *h*. They have a mean annual soil temperature lower than 22°C, and the mean summer and mean winter soil temperatures at a depth of 50 cm differ by 5°C or more. They have an aridic moisture regime that borders on a xeric regime.

Xerertic Haplargids are like Typic Haplargids except for *g* or *e* and *g* and have a thermic, mesic, or frigid soil temperature regime, and they have cracks that close for 60 consecutive days or more during the 120 days following the winter solstice in more than 7 out of 10 years.

Xerollic Haplargids are like Typic Haplargids except for *e* with or without *h*, have a mean annual soil temperature lower than 22°C, and the mean summer and mean winter soil temperatures at a depth of 50 cm differ by 5°C or more. They have an aridic moisture regime that borders on a xeric regime.

Nadurargids

Distinctions between Typic Nadurargids and other subgroups

Typic Nadurargids are the Nadurargids that
a. Are not saturated with water in any horizon within a depth of 1 m at any period or do not have either of the following characteristics within the horizon or horizons that are saturated:
    (1) Dominant chroma of 1 or less throughout and hue as yellow or yellower than 2.5Y in some part;
    (2) Both a dominant chroma of 2 or less and mottles that are not due to segregated lime;
b. Have a platy or massive duripan that is indurated in some subhorizon; and
c. Have a duripan shallower than 18 cm; or have a weighted average percentage of organic carbon in the upper soil to a depth of 40 cm that is <0.6 if the weighted average ratio of sand to clay to that depth is 1.0 or less, or that is less than one-seventh percent if the ratio of sand to clay is 13 or more, or have an intermediate amount of organic carbon if the ratio of sand to clay is between 1.0 and 13; or have a weighted average percentage of organic carbon in the surface soil to a depth of 18 cm that is not one-fifth more than the values just stated if there is a duripan that is shallower than 40 cm but deeper than 18 cm.
Aquic Haplic Nadurargids are like Typic Nadurargids except for *a* and *b*, with or without *c*.
Haplic Nadurargids are like Typic Nadurargids except for *b*.
Haploxerollic Nadurargids are like Typic Nadurargids except for *b* and *c*. They have an aridic moisture regime that borders on a xeric regime.
Xerollic Nadurargids are like Typic Nadurargids except for *c*. They have a mean annual soil temperature lower than 22°C, and the mean summer and mean winter soil temperatures at a depth of 50 cm differ by 5°C or more. They have an aridic moisture regime that borders on a xeric regime.

Natrargids

Distinctions between Typic Natrargids and other subgroups

Typic Natrargids are the Natrargids that
a. Either are not saturated with water in any horizon within 1 m of the surface at any time or do not have either of the following characteristics in the horizon or horizons that are saturated:
    (1) Dominant chroma of 1 or less throughout and hue as yellow or yellower than 2.5Y in some part; or
    (2) Both a dominant chroma of 2 or less accompanied by mottles that are not due to segregated lime;
b. Do not have a horizon within 1 m of surface that is more than 15 cm thick and that either contains 20 percent or

more durinodes in a nonbrittle matrix or is brittle and has firm consistence when moist;

c. Do not have more than 10 percent of the ped surfaces deeper than 2.5 cm below the upper boundary of the natric horizon covered by skeletans;

d. Do not have a lithic contact within 50 cm of the soil surface;

e. Have a weighted average percentage of organic carbon in the upper soil to a depth of 40 cm that is <0.6 if the weighted average of sand to clay in that depth is 1.0 or less, or is not more than one-seventh percent if the ratio is 13 or more, or has an intermediate percentage of organic carbon if the ratio is between 1.0 and 13; or have a weighted average percentage of organic carbon in the surface soil to a depth of 18 cm that is not one-fifth more than the values just stated if there is a lithic or paralithic contact that is shallower than 40 cm but deeper than 18 cm; and

f. Have an SAR of >13 or have 15 percent or more saturation with sodium throughout the major part of the natric horizon.

Aquic Natrargids are like Typic Natrargids except for *a* or for *a* and *e*.

Borollic Natrargids are like Typic Natrargids except for *e* or for *e* and *f*. The mean annual soil temperature is lower than 8°C, the mean summer and mean winter soil temperatures at a depth of 50 cm differ by >5°C, and the moisture regime is aridic bordering on ustic.

Borollic Glossic Natrargids are like Typic Natrargids except for *c* and *e*, with or without *f*. The mean annual soil temperature is lower than 8°C, the mean summer and mean winter soil temperatures at a depth of 50 cm differ by >5°C, and the moisture regime is aridic bordering on ustic.

Duric Natrargids are like Typic Natrargids except for *b* or for *b* and *f*.

Durixerollic Natrargids are like Typic Natrargids except for *b* and *e*. They have a mean annual soil temperature lower than 22°C. The mean summer and mean winter soil temperatures at a depth of 50 cm differ by 5°C or more, and the moisture regime is aridic bordering on xeric.

Glossic Ustollic Natrargids are like Typic Natrargids except for *c* and *e*, with or without *f*. They have a mean annual soil temperature of 8°C or higher and an aridic moisture regime that borders on ustic.

Haplic Natrargids are like Typic Natrargids except for *f*.

Haploxerollic Natrargids are like Typic Natrargids except for *e* and *f*. They have a mean annual soil temperature lower than 22°C, and the mean summer and mean winter soil temperatures at a depth of 50 cm differ by 5°C or more. They have an aridic moisture regime that borders on a xeric regime.

Haplustollic Natrargids are like Typic Natrargids except for *e* and f. They have a mean annual soil temperature of 8°C or higher, and the moisture regime is aridic but borders on ustic.

Lithic Natrargids are like Typic Natrargids except for *d*.

Lithic Xerollic Natrargids are like Typic Natrargids except for *d* and *e*. They have a mean annual soil temperature lower than 22°C, and the mean summer and mean winter soil temperatures at a depth of 50 cm differ by 5°C or more. They have an aridic moisture regime that borders on a xeric regime.

Ustollic Natrargids are like Typic Natrargids except for *e*.
They have a mean annual soil temperature of 8°C or higher
and an aridic moisture regime that borders on an ustic
regime.
Xerollic Natrargids are like Typic Natrargids except for *e*.
They have a mean annual soil temperature lower than 22°C,
and the mean summer and mean winter soil temperatures at
a depth of 50 cm differ by 5°C or more. The moisture
regime is aridic, bordering on xeric.

Paleargids

Distinctions between Typic Paleargids and other subgroups

Typic Paleargids are the Paleargids that
a. Do not have a horizon within 1 m of the surface that is
>15 cm thick that either contains 20 percent or more
durinodes in a nonbrittle matrix or is brittle and has firm
consistence when moist;
b. Have a weighted average percentage of organic carbon in
the upper soil to a depth of 40 cm of <0.6 percent if the
weighted average ratio of sand to clay above this depth is
1.0 or less, or one-seventh percent if the ratio is 13 or more,
or an intermediate percentage of organic carbon if the ratio
of sand to clay is between 1.0 and 13; or a weighted average
percentage of organic carbon in the surface soil to a depth
of 18 cm that is not one-fifth more than the values just
stated if there is a petrocalcic horizon whose upper
boundary is shallower than 40 cm but deeper than 18 cm;
c. Do not have a petrocalcic horizon whose upper boundary
is within 1 m of the soil surface;
d. Have either
　(1) An increase of 15 percent or more clay (absolute)
　within a vertical distance of 2.5 cm at the upper
　boundary of the argillic horizon, or
　(2) An increase of 10 percent or more clay (absolute) if
　the soil is cultivated and the lower boundary of the Ap
　horizon is the upper boundary of the argillic horizon;
e. Do not have the following combination of characteristics:
　(1) Cracks at some period in most years that are 1 cm or
　more wide at a depth of 50 cm, that are at least 30 cm
　long in some part, and that extend upward to the soil
　surface, to the base of an Ap horizon, or to the top of
　the argillic horizon;
　(2) A coefficient of linear extensibility (COLE) of 0.05
　or more in a horizon or horizons at least 50 cm thick and
　a potential linear extensibility of 6 cm or more in the
　upper 1.5 m of the soil or in the whole soil if a lithic or
　paralithic contact is deeper than 50 cm but shallower than
　1.5 m; and
　(3) More than 35 percent clay in horizons that total >50
　cm in thickness; and
f. Are dry in all parts of the moisture control section more
than three-fourths of the time (cumulative) that the soil
temperature at a depth of 50 cm is 5°C or higher.
Borollic Paleargids are like Typic Paleargids except for *b* or
for *b* and *f*. The mean annual soil temperature is lower than
8°C, and the mean summer and mean winter soil
temperatures at a depth of 50 cm differ by >5°C. They
have an aridic moisture regime that borders on an ustic
regime.

Borollic Vertic Paleargids are like Typic Paleargids except
for *b* and *e*, with or without *f*. The mean annual soil
temperature is lower than 8°C, the mean summer and mean
winter soil temperatures at a depth of 50 cm differ by >5°C,
and the cracks are not closed for as many as 60 consecutive
days of the 120 days following the winter solstice in 3 or
more years out of 10.
Petrocalcic Paleargids are like Typic Paleargids except for *c*
or for *c* and *d*.
Petrocalcic Ustalfic Paleargids are like Typic Paleargids
except for *c* and *f*, with or without *d*. They have a mean
annual soil temperature that is 8°C or higher and an aridic
moisture regime that borders on an ustic regime.
Petrocalcic Ustollic Paleargids are like Typic Paleargids
except for *b* and *c*, with or without *d* or *f*, or both. They
have a mean annual soil temperature that is 8°C or higher
and an aridic moisture regime that borders on an ustic
regime.
Petrocalcic Xerollic Paleargids are like Typic Paleargids
except for *b* and *c*, with or without *d* or *f*, or both. They
have a mean annual soil temperature lower than 22°C, and
the mean summer and mean winter soil temperatures at a
depth of 50 cm differ by 5°C or more. They have an aridic
moisture regime that borders on a xeric regime.
Ustollic Paleargids are like Typic Paleargids except for *b* or
for *b* and *f*. They have a mean annual soil temperature that
is 8°C or higher and an aridic moisture regime that borders
on an ustic regime.
Xeralfic Paleargids are like Typic Paleargids except for *f*.
They have a mean annual soil temperature lower than 22°C,
and the mean summer and mean winter soil temperatures at
a depth of 50 cm differ by 5°C or more. They have an
aridic moisture regime that borders on a xeric regime.
Xerollic Paleargids are like Typic Paleargids except for *b*,
with or without *f*. They have a mean annual soil
temperature lower than 22°C, and the mean summer and
mean winter soil temperatures at a depth of 50 cm differ by
5°C or more. They have an aridic moisture regime that
borders on a xeric regime.

Orthids

Key to great groups

EBA. Orthids that have a salic horizon whose upper
boundary is within 75 cm of the soil surface and are
saturated with water within a depth of 1 m for 1 month or
more in most years or have artificial drainage and lack a
duripan that has an upper boundary within 1 m of the soil
surface.

Salorthids, p. 109

EBB. Other Orthids that have a petrocalcic horizon whose
upper boundary is within 1 m of the soil surface and is not
overlain by a duripan.

Paleorthids, p. 108

EBC. Other Orthids that have a duripan whose upper
boundary is within 1 m of the soil surface.

Durorthids, p. 107

EBD. Other Orthids that have a gypsic or petrogypsic horizon whose upper boundary is within 1 m of the soil surface.

Gypsiorthids, p. 107

EBE. Other Orthids that have a calcic horizon whose upper boundary is within 1 m of the surface and that are calcareous in all parts above the calcic horizon after the upper soil, to a depth of 18 cm, is mixed unless the texture is as coarse or coarser than loamy fine sand.

Calciorthids, p. 103

EBF. Other Orthids (that have a cambic horizon).

Camborthids, p. 104

Calciorthids

Distinctions between Typic Calciorthids and other subgroups

Typic Calciorthids are the Calciorthids that
a. Are not saturated with water for 90 consecutive days or more in most years within 1 m of the surface and do not have any of the following characteristics within a depth of 1 m below the surface if the soil above that depth is saturated with water at some period in most years or the soil is artificially drained:
  (1) Dominant chroma of 1 or less throughout and hue as yellow or yellower than 2.5Y in some part;
  (2) Dominant chroma of 2 or less and mottles that are not due to segregated lime; or
  (3) Both a dominant chroma of 2 or less and a greater SAR (or percentage of exchangeable Na) in more than half the thickness of the horizons between the surface and 50 cm than in the saturated zone;
b. Do not have a horizon within 1 m of the surface that is >15 cm thick that either contains 20 percent or more durinodes in a nonbrittle matrix or is brittle and has firm consistence when moist;
c. Do not have a lithic contact within 50 cm of the surface;
d. Have a weighted average content of organic carbon in the upper soil to a depth of 40 cm that is <0.6 percent if the weighted average ratio of sand to noncarbonate clay for this depth is 1.0 or less, or is <0.15 percent if the ratio is 13 or more, or is intermediate if the ratio of sand to clay is between 1.0 and 13; or a weighted average percentage of organic carbon in the surface soil to a depth of 18 cm that is not one-fifth more than the values just stated if there is a lithic or paralithic contact that is shallower than 40 cm but deeper than 18 cm;
e. Are dry in all parts of the moisture control section more than three-fourths of the time (cumulative) that the soil temperature is 5°C or higher at a depth of 50 cm; and
f. Do not have reddish peds below the calcic horizon that are weakly calcareous or noncalcareous but that are thickly coated with lime.
Aquic Calciorthids are like Typic Calciorthids except for *a*, with or without *d* or *e*, or both.
Aquic Duric Calciorthids are like Typic Calciorthids except for *a* and *b*, with or without *d* or *e*, or both.
Argic Calciorthids are like Typic Calciorthids except for *f*.

Borollic Calciorthids are like Typic Calciorthids except for *d* or for *d* and *e*. The soil temperature regime is frigid, and the moisture regime is aridic but borders on ustic.

Borollic Lithic Calciorthids are like Typic Calciorthids except for *c* and *d*, with or without *e*. The soil temperature regime is frigid, and the moisture regime is aridic but borders on ustic.

Durixerollic Calciorthids are like Typic Calciorthids except for *b* and *d*, with or without *e*, and have a mean annual soil temperature lower than 22°C. The mean summer and mean winter soil temperatures at a depth of 50 cm differ by 5°C or more, and the moisture regime is aridic but borders on xeric.

Lithic Calciorthids are like Typic Calciorthids except for *c*.

Lithic Ustollic Calciorthids are like Typic Calciorthids except for *c* and *d*, with or without *e*. They have a mean annual soil temperature that is 8°C or higher and an aridic moisture regime that borders on an ustic regime.

Lithic Xerollic Calciorthids are like Typic Calciorthids except for *c* and *d*, with or without *e*. They have a mean annual soil temperature lower than 22°C, the mean summer and mean winter soil temperatures at a depth of 50 cm differ by 5°C or more, and they have an aridic moisture regime that borders on a xeric regime.

Ustochreptic Calciorthids are like Typic Calciorthids except for *e*. They have a mean annual soil temperature that is 8°C or higher and an aridic moisture regime that borders on an ustic moisture regime.

Ustollic Calciorthids are like Typic Calciorthids except for *d* or for *d* and *e*. They have a mean annual soil temperature that is 8°C or higher and an aridic moisture regime that borders on an ustic regime.

Xerollic Calciorthids are like Typic Calciorthids except for *d* or for *d* and *e*. They have a mean annual soil temperature lower than 22°C, the mean summer and mean winter soil temperatures at a depth of 50 cm differ by 5°C or more, and they have an aridic moisture regime that borders on a xeric regime.

Camborthids

Distinctions between Typic Camborthids and other subgroups

Typic Camborthids are the Camborthids that
a. Are not saturated with water for 90 consecutive days or more within 1 m of the surface in most years and do not have any of the following characteristics within 1 m of the soil surface if the soil of that zone is saturated with water at some period in most years or the soil is artificially drained:

(1) Dominant chroma of 1 or less throughout and hue as yellow or yellower than 2.5Y in some part,

(2) Dominant chroma of 2 or less and mottles that are due to segregation of iron or manganese, or

(3) Both a dominant chroma of 2 or less and a greater SAR (or percentage of exchangeable Na) in more than half the thickness of the horizons between the surface and a depth of 50 cm than in the saturated zone;

b. Do not have a horizon within 1 m of the surface that is >15 cm thick that either contains 20 percent or more durinodes in a nonbrittle matrix or is brittle and has firm consistence when moist;

c. Do not have a lithic contact within 50 cm of the surface;

d. Have a weighted average organic-carbon content in the upper soil to a depth of 40 cm that is <0.6 percent if the weighted average ratio of sand to noncarbonate clay to this depth is 1.0 or less, or is <0.15 percent if the ratio is 13 or more, or an intermediate weighted average percentage of organic carbon if the ratio is between 1.0 and 13; or a weighted average percentage of organic carbon in the upper soil to a depth of 18 cm that is not as much as one-fifth more than the values just stated if there is a lithic or paralithic contact shallower than 40 cm but deeper than 18 cm;

e. Are dry in all parts of the moisture control section for more than three-fourths of the time (cumulative) that the soil temperature is 5°C or more at a depth of 50 cm unless the soil is irrigated;

f. Do not have the following combination of characteristics:

(1) Cracks at some period in most years that are 1 cm or more wide at a depth of 50 cm, that are at least 30 cm long in some part, and that extend upward to the soil surface or to the base of an Ap horizon,

(2) A coefficient of linear extensibility (COLE) of 0.05 or more in a horizon or horizons at least 50 cm thick and a potential linear extensibility of 6 cm or more in the upper 1.5 m of the soil or in the whole soil if a lithic or paralithic contact is deeper than 50 cm but shallower than 1.5 m, and

(3) More than 35 percent clay in horizons that total >50 cm in thickness;

g. Have a content of organic carbon that decreases regularly with depth below a depth of 25 cm and, unless a lithic or paralithic contact occurs at a shallower depth, reaches a level of <0.2 percent at a depth 1.25 m below the surface;

h. Have an SAR of 45 or less or <40 percent saturation with sodium throughout the cambic horizon if the saturated hydraulic conductivity is slow or very slow; and

i. Do not have an anthropic epipedon.

Anthropic Camborthids are like Typic Camborthids except for *d* and *i*.

Aquic Camborthids are like Typic Camborthids except for *a*, with or without *d* or *e*, or both.

Aquic Duric Camborthids are like Typic Camborthids except for *a* and b, with or without *d* or *e*, or both.

Borollic Camborthids are like Typic Camborthids except for *d* or for *d* and *e*. The mean annual soil temperature is lower than 8°C, the mean summer and mean winter soil temperatures at a depth of 50 cm differ by >5°C, and the moisture regime is aridic bordering on an ustic regime.

Borollic Lithic Camborthids are like Typic Camborthids except for *c* and *d*, with or without *e*. The mean annual soil temperature is lower than 8°C, the mean summer and mean winter soil temperatures at a depth of 50 cm differ by >5°C, and the moisture regime is aridic bordering on ustic.

Borollic Vertic Camborthids are like Typic Camborthids except for *d* and *f*, with or without *e*. The mean annual soil temperature is lower than 8°C, and the mean summer and mean winter soil temperatures at a depth of 50 cm differ by >5°C. The cracks are not closed for as many as 60 consecutive days of the 120 days following the winter solstice in 3 or more years out of 10.

Duric Camborthids are like Typic Camborthids except for *b*.

Durixerollic Camborthids are like Typic Camborthids except for *b* and *d*, with or without *e*. They have a mean annual soil temperature lower than 22°C, and the mean summer and mean winter soil temperatures at a depth of 50 cm differ by 5°C or more. They have an aridic moisture regime that borders on a xeric regime.

Durixerollic Lithic Camborthids are like Typic Camborthids except for *b*, *c*, and *d*, with or without *e*. They have a mean annual soil temperature lower than 22°C, and the mean summer and mean winter soil temperatures at a depth of 50 cm differ by 5°C or more. They have an aridic moisture regime that borders on a xeric regime.

Fluventic Camborthids are like Typic Camborthids except for *g*.

Lithic Camborthids are like Typic Camborthids except for *c*.

Lithic Xerollic Camborthids are like Typic Camborthids except for *c* and *d*, with or without *e*. They have a mean annual soil temperature lower than 22°C, and the mean summer and mean winter soil temperatures at a depth of 50 cm differ by 5°C or more. They have an aridic moisture regime that borders on a xeric regime.

Natric Camborthids are like Typic Camborthids except for *h*, with or without any or all of *a*, *d*, *e*, *f*, and *g*, and they have slow or very slow saturated hydraulic conductivity.

Ustertic Camborthids are like Typic Camborthids except for *f*, with or without any or all of *d*, *e*, and *g*. In most years, unless irrigated, they have cracks that remain open for 175 to 240 days, cumulative. The cracks are not closed for as many as 60 consecutive days during the 120 days following the winter solstice in 3 or more years out of 10 if the soil temperature regime is thermic, mesic, or frigid.

Ustochreptic Camborthids are like Typic Camborthids except for *e*. They have an aridic moisture regime that borders on an ustic moisture regime and a hyperthermic, thermic, or mesic soil temperature regime.

Ustollic Camborthids are like Typic Camborthids except for *d* or for *d* and *e*, have a mean annual soil temperature that is 8°C or higher, and have an aridic moisture regime that borders on an ustic regime.

Vertic Camborthids are like Typic Camborthids except for *f* or for *e* and *f*. Unless the soils are irrigated, the cracks remain open in most years for more than 240 days, cumulative, and are not closed in most years for as many as 60 consecutive days at any season.

Xerertic Camborthids are like Typic Camborthids except for *f*, with or without any or all of *d*, *e*, and *g*. They have a thermic, mesic, or frigid soil temperature regime and have cracks that are closed for 60 consecutive days or more during the 120 days following the winter solstice in more than 7 years out of 10.

Xerollic Camborthids are like Typic Camborthids except for *d* or for *d* and *e*. They have a mean annual soil temperature lower than 22°C, and the mean summer and mean winter soil temperatures at a depth of 50 cm differ by 5°C or more. They have an aridic moisture regime that borders on a xeric regime.

Durorthids

Distinctions between Typic Durorthids and other subgroups

Typic Durorthids are the Durorthids that
a. Are not saturated with water for 90 consecutive days or more in most years within 1 m of the surface and do not have a subhorizon within 1 m of the soil surface that has the following characteristics if the horizon is saturated with water at some period in most years or the soil is artificially drained:
    (1) Dominant chroma of 1 or less throughout and hue as yellow or yellower than 2.5Y in some part;
    (2) Dominant chroma of 2 or less accompanied by mottles that are not due to segregated lime; or
    (3) Both a dominant chroma of 2 or less and a greater SAR (or percentage of exchangeable sodium) in more than half the thickness of the horizons between the surface and 50 cm depth than in the saturated zone;
b. Have a platy or massive duripan that is indurated in some subhorizon;
c. Have a duripan whose upper boundary is shallower than 18 cm; or have a weighted average content of organic carbon in the upper soil to a depth of 40 cm that is <0.6 if the weighted average ratio of sand to noncarbonate clay above that depth is 1.0 or less or is <0.15 percent if the ratio is 13 or more, or is intermediate if the ratio of sand to clay is between 1.0 and 13;
or a weighted average percentage of organic carbon in the surface soil to a depth of 18 cm that is not one-fifth more than the values just stated if there is a duripan shallower than 40 cm but deeper than 18 cm; and
d. Are dry in all parts of the moisture control section for more than three-fourths of the time that the soil temperature at a depth of 50 cm is 5°C or higher.
Aquentic Durorthids are like Typic Durorthids except for *a* and *b*, with or without *c* or *d*, or both.
Aquic Durorthids are like Typic Durorthids except for *a*, with or without *c* or *d*, or both.
Entic Durorthids are like Typic Durorthids except for *b*.
Haploxerollic Durorthids are like Typic Durorthids except for *b* and *c*, with or without *d*. They have a mean annual soil temperature lower than 22°C, and the mean summer and mean winter soil temperatures at a depth of 50 cm differ by 5°C or more. They have an aridic moisture regime that borders on a xeric regime.
Xerollic Durorthids are like Typic Durorthids except for *c* or for *c* and *d*. They have a mean annual soil temperature lower than 22°C, and the mean summer and mean winter soil temperatures at a depth of 50 cm differ by 5°C or more. They have an aridic moisture regime that borders on a xeric regime.

Gypsiorthids

Distinctions between Typic Gypsiorthids and other subgroups

Typic Gypsiorthids are the Gypsiorthids that
a. Do not have a petrogypsic horizon whose upper boundary is within 1 m of the soil surface; and

b. Have a gypsic horizon in which the product of the percentage of gypsum and the thickness in centimeters above a depth of 1.5 m is 3,000 or more.

Calcic Gypsiorthids are like Typic Gypsiorthids except for *b* and have a calcic horizon above the gypsic horizon.

Cambic Gypsiorthids are like Typic Gypsiorthids except for *b* and do not have a calcic horizon above the gypsic horizon.

Petrogypsic Gypsiorthids are like Typic Gypsiorthids except for *a* with or without *b*.

Paleorthids

Distinctions between Typic Paleorthids and other subgroups

Typic Paleorthids are the Paleorthids that
a. Are not saturated with water for 90 consecutive days or more in most years within 1 m below the soil surface and do not have a subhorizon that has the following characteristics within 1 m of the surface if the horizon is saturated with water at some period in most years or the soil is artificially drained. The soils do not have:
> (1) Dominant chroma of 1 or less throughout and hue as yellow or yellower than 2.5Y in some part;
> (2) Dominant chroma of 2 or less and mottles that are not due to segregated lime; or
> (3) Both a dominant chroma of 2 or less and a greater SAR (or percentage of exchangeable Na) in more than half the thickness of the horizons between the surface and 50 cm depth than in the saturated zone;

b. Have a petrocalcic horizon whose upper boundary is shallower than 18 cm, or have a weighted average content of organic carbon in the upper soil to a depth of 40 cm that is <0.6 percent if the weighted average ratio of sand to non-carbonate clay to this depth is 1.0 or less, or is <0.15 percent if the ratio is 13 or more, or is intermediate if the ratio of sand to clay is between 1.0 and 13; or have a weighted average percentage of organic carbon in the surface soil to a depth of 18 cm that is not one-fifth more than the values just stated if there is a petrocalcic horizon that is shallower than 40 cm but deeper than 18 cm; and

c. Are dry in all parts of the moisture control section for more than three-fourths of the time (cumulative) that the soil temperature is 5°C or higher at a depth of 50 cm.

Aquic Paleorthids are like Typic Paleorthids except for *a*, with or without *b* or *c*, or both.

Ustochreptic Paleorthids are like Typic Paleorthids except for *c*. They have a mesic, thermic, or hyperthermic soil temperature regime and an aridic moisture regime that borders on an ustic regime.

Ustollic Paleorthids are like Typic Paleorthids except for *b* or for *b* and *c*. They have a mean annual soil temperature that is 8°C or higher and an aridic moisture regime that borders on an ustic regime.

Xerollic Paleorthids are like Typic Paleorthids except for *b* or for *b* and *c*. They have a mean annual soil temperature lower than 22°C, and the mean summer and mean winter soil temperatures at a depth of 50 cm differ by 5°C or more. They have an aridic moisture regime that borders on a xeric regime.

Salorthids

Distinctions between Typic Salorthids and Aquollic Salorthids

Typic Salorthids are the Salorthids that
a. Have a salic horizon that has its upper boundary within 18 cm of the soil surface; and
b. Have a weighted average content of organic carbon in the upper soil to a depth of 40 cm that is <0.6 percent if the weighted average ratio of sand to noncarbonate clay to this depth is 1.0 or less, or is <0.15 percent if the ratio is 13 or more, or is intermediate if the ratio of sand to clay is between 1.0 and 13.
Aquollic Salorthids are like Typic Salorthids except for *b*.

Chapter 7

Entisols

Key To Suborders

JA. Entisols that
    1. Have sulfidic materials within 50 cm of the mineral soil surface; or
    2. Are permanently saturated with water and have in all horizons below 25 cm
        a. Dominant hue that is neutral or bluer than 10Y and
        b. Colors that change on exposure to the air; or
    3. Are saturated with water at some time of year or are artificially drained and have, within 50 cm of the surface, dominant color (moist) in the matrix as follows:
        a. In horizons that have texture finer than loamy fine sand in some or all subhorizons, or that have >35 percent (by volume) of rock fragments in some subhorizon
            (1) If there is mottling, chroma is 2 or less;
            (2) If there is no mottling and the value is less than 4, chroma is less than 1; if the value is 4 or more, chroma is 1 or less;
        b. In horizons that have texture of loamy fine sand or coarser in all subhorizons
            (1) If the hue is as red or redder than 10YR and there is mottling, chroma is 2 or less; if there is no mottling and the value is less than 4, chroma is less than 1; or if the value is 4 or more, chroma is 1 or less;
            (2) If the hue is between 10YR and 10Y and there is distinct or prominent mottling, chroma is 3 or less; if there is no mottling, chroma is 1 or less;
            (3) Hue is bluer than 10Y; or
            (4) Any color if the color is due to uncoated grains of sand.

Aquents, p. 112

JB. Other Entisols that have fragments of diagnostic horizons between 25 cm and 1 m below the surface, but the fragments are not arranged in discernible order.

Arents, p. 116

JC. Other Entisols that have below the Ap horizon or below a depth of 25 cm, whichever is deeper, <35 percent (by volume) of rock fragments and that have texture of loamy fine sand or coarser in all subhorizons[1] either to a depth of 1 m or to a lithic, paralithic, or petroferric contact, whichever is shallower.

Psamments, p. 125

JD. Other Entisols that do not have a lithic or paralithic contact within 25 cm of the soil surface and that have slopes of <25 percent and organic carbon content that decreases irregularly with depth or remains above a level of 0.2

---

[1]Lamellae that are <1 cm thick or that are too few to meet the requirements for an argillic horizon are permitted to have texture of sandy loam. See the definition of an argillic horizon (ch. 1).

percent to a depth of 1.25 m, and the mean annual soil temperature is higher than 0°C. (Strata of sand or loamy sand may have less organic carbon if finer sediments at a depth of 1.25 m or below have 0.2 percent organic carbon or more).

Fluvents, p. 117

JE. Other Entisols.

Orthents, p. 120

Aquents

Key to great groups

JAA. Aquents that have sulfidic materials within 50 cm of the mineral soil surface.

Sulfaquents, p. 116

JAB. Other Aquents that have an n value of >0.7 and that have at least 8 percent clay in all subhorizons between a depth of 20 and 50 cm and that have a mean annual soil temperature higher than 0°C.

Hydraquents, p. 115

JAC. Other Aquents that have a cryic but not a pergelic[2] soil temperature regime.

Cryaquents, p. 113

JAD. Other Aquents that have an organic carbon content[3] that decreases irregularly with depth or that remains above 0.2 percent to a depth of 1.25 m; and that have texture finer than loamy fine sand in some or all subhorizons between the Ap horizon or a depth of 25 cm, whichever is deeper, and 1 m or a lithic or paralithic contact, whichever is shallower. Thin strata of sand may have less organic carbon if the finer sediments at a depth of 1.25 m or below have 0.2 percent organic carbon or more.

Fluvaquents, p. 113

JAE. Other Aquents that have a difference of <5°C between the mean summer and mean winter soil temperatures at a depth of 50 cm.

Tropaquents, p. 116

JAF. Other Aquents that have a sandy particle-size class in all subhorizons between the Ap horizon or a depth of 25 cm, whichever is deeper, and 1 m or a lithic or paralithic contact, whichever is shallower, and that have mean summer and mean winter soil temperatures at a depth of 50 cm that differ by 5°C or more.

Psammaquents, p. 115

JAG. Other Aquents.

Haplaquents, p. 114

---

[2] Soils that otherwise could be Aquents are grouped with Aquepts if there is permafrost.

[3] The carbon should be of Holocene age. It is not the intent to include fossil carbon from transported fragments of bedrock or from buried Pleistocene deposits. The mean residence time of the carbon should be <11,000 years B.P.

Cryaquents

Distinctions between Typic Cryaquents and other subgroups

Typic Cryaquents are the Cryaquents that
a. Do not have a layer in the upper 75 cm that has texture finer than loamy fine sand, that is as much as 18 cm thick, that has a bulk density of the fine-earth fraction (at 33 kPa moisture tension) of 0.95 g per cubic centimeter or less, and that has either
   (1) A ratio of measured clay to 1500 kPa water (percentages) of 1.25 or less, or
   (2) A ratio of CEC (at pH near 8) to 1500 kPa water of more than 1.5 and more exchange acidity than the sum of bases plus KCl-extractable aluminum.
Andaqueptic Cryaquents are like Typic Cryaquents except for *a.*

Fluvaquents

Distinctions between Typic Fluvaquents and other subgroups

Typic Fluvaquents are the Fluvaquents that
a. Have, in 60 percent or more of the matrix in all subhorizons between the Ap horizon or a depth of 25 cm, whichever is deeper, and a depth of 75 cm, one or more of the following:
   (1) If mottled and
      (a) If the hue is 2.5Y or redder[4] and the value, moist, is more than 5, the chroma, moist, is 2 or less;
      (b) If the hue is 2.5Y or redder and the value, moist, is 5 or less, the chroma, moist, is 1 or less; or
      (c) If the hue is yellower than 2.5Y, the chroma, moist, is 2 or less; or
   (2) The chroma, moist, is 1 or less and mottles may or may not be present;
b. Do not have a layer in the upper 75 cm that has texture finer than loamy fine sand, that is as much as 18 cm thick, that has a bulk density in the fine-earth fraction (at 33 kPa moisture tension) of 0.95 grams per cubic centimeter or less, and that has either (1) a ratio of measured clay to 1500 kPa water (percentages) of 1.25 or less or (2) a ratio of CEC (at pH near 8) to 1500 kPa water of >1.5 and more exchange acidity than the sum of bases plus KCl-extractable aluminum;
c. Do not have the following combination of characteristics:
   (1) Cracks at some period in most years that are 1 cm or more wide at a depth of 50 cm, that are at least 30 cm long in some part, and that extend upward to the surface or to the base of an Ap horizon;
   (2) A coefficient of linear extensibility (COLE) of 0.09 or more in a horizon or horizons at least 50 cm thick and a potential linear extensibility of 6 cm or more in the upper 1 m of the soil or in the whole soil if a lithic or

---

[4]If the hue of the matrix is 7.5YR or redder and if peds are present, ped exteriors have dominant chroma, moist, of 1 or less, and ped interiors have mottles that have chroma, moist, of 2 or less; if peds are absent, the chroma, moist, is 1 or less immediately below any surface horizon that has value, moist, less than 3.5.

paralithic contact is deeper than 50 cm but shallower than 1 m; and

(3) More than 35 percent clay in horizons that total >50 cm in thickness;

d. Have an Ap horizon that has a color value, moist, of 4 or more or has a color value, dry, of 6 or more when crushed and smoothed, or the A horizon is <15 cm thick if its color value, moist, is <3.5;

e. Do not have a buried Histosol or a buried histic epipedon that has its upper boundary within 1 m of the soil surface;

f. Do not have sulfidic materials within 1 m of the mineral soil surface; and

g. Have a difference of 5°C or more between the mean summer and mean winter soil temperatures at a depth of 50 cm or at a lithic or paralithic contact, whichever is shallower.

Aeric Fluvaquents are like Typic Fluvaquents except for *a* or *a* and *d*.

Aeric Tropic Fluvaquents are like Typic Fluvaquents except for *a* and *g*, with or without *d*.

Andaqueptic Fluvaquents are like Typic Fluvaquents except for *b*.

Humaqueptic Fluvaquents are like Typic Fluvaquents except for *d*, and the base saturation (by NH_4OAc) is <50 percent in some horizon and does not increase to 50 percent or more within a depth of 1 m below the soil surface.

Mollic Fluvaquents are like Typic Fluvaquents except for *d*, and the base saturation (by NH_4OAc) is 50 percent or more throughout the soil or increases to 50 percent or more within a depth of 1 m below the soil surface.

Sulfic Fluvaquents are like Typic Fluvaquents except for *f*.

Thapto-Histic Fluvaquents are like Typic Fluvaquents except for *e* with or without *a* or *d* or both.

Thapto-Histic Tropic Fluvaquents are like Typic Fluvaquents except for *e* and *g*, with or without *d*.

Tropic Fluvaquents are like Typic Fluvaquents except for *g* or for *g* and *d*.

Vertic Fluvaquents are like Typic Fluvaquents except for *c*, with or without *a* or *d*, or both, and the cracks are not open permanently.

Haplaquents

Distinctions between Typic Haplaquents and other subgroups

Typic Haplaquents are the Haplaquents that

a. Have in 60 percent or more of the matrix in all subhorizons between the Ap horizon or a depth of 25 cm, whichever is deeper, and 75 cm one or more of the following:

(1) If mottled and

(a) If the hue is 2.5Y or redder[5] and the value, moist, is more than 5, the chroma, moist, is 2 or less;

(b) If the hue is 2.5Y or redder and the value, moist, is 5 or less, the chroma, moist, is 1 or less;

---

[5] If the hue of the matrix is 7.5YR or redder and if peds are present, ped exteriors have dominant chroma, moist, of 1 or less, and ped interiors have mottles that have chroma, moist, of 2 or less; if peds are absent, the chroma, moist, is 1 or less immediately below any surface horizon that has value, moist, less than 3.5.

(c) If the hue is yellower than 2.5Y, the chroma,
moist, is 2 or less; or
(2) The chroma, moist, is 1 or less and mottles may or
may not be present;
b. Have an Ap horizon that has a color value, moist, of 4 or
more or that has a color value, dry, of 6 or more when
crushed and smoothed, or the A1 horizon is <15 cm thick if
its color value, moist, is less than 3.5;
c. Do not have a lithic contact within 50 cm of the soil
surface; and
d. Do not have sulfidic materials within 1 m of the mineral
soil surface.
Aeric Haplaquents are like Typic Haplaquents except for *a*
or *a* and *b*.
Mollic Haplaquents are like Typic Haplaquents except for *b*.
Sulfic Haplaquents are like Typic Haplaquents except for *d*.

Hydraquents

Hydraquents are the Aquents that do not have sulfidic
materials within 50 cm of the mineral soil surface and
1. Have a mean annual soil temperature higher than 0°C;
2. In all subhorizons between 20 and 50 cm below the
mineral surface have both an n value of more than 0.7 and
at least 8 percent clay; and
3. Have texture that is loamy very fine sand or finer in
some horizon below the Ap horizon or a depth of 25 cm,
whichever is deeper, but above a depth of 1 m or a lithic or
a paralithic contact, whichever is shallower.

The Hydraquents have been little used or studied in the
United States, and subgroups have not been developed. The
typic subgroup could be defined as having a high n value, 1
or more, in all subhorizons between a depth of 20 cm and 1
m. Hydraquents that have n values <1 in the upper 25 cm
or more could be Haplic Hydraquents, which are intergrades
to Haplaquents. Thapto-Histic Hydraquents, which have a
buried Histosol or histic epipedon whose upper boundary is
within a depth of 1 m, doubtless exist. Hydraquents that
have a small amount of sulfides or none in the upper
horizons can be drained and cultivated. A sulfic subgroup
might be useful for those soils that, after drainage, have a
pH <4.5 (1:1 water) in the upper 25 cm or more, those that
have enough sulfides in the undrained soil to produce this
pH on drainage, and those that have a larger amount of
sulfides at a depth between 50 cm and 1 m. A large amount
of sulfides at a depth below 1 m should be noted but could
be shown as a phase.

Psammaquents

Distinctions between Typic Psammaquents and other
subgroups

Typic Psammaquents are the Psammaquents that
a. Do not have a lithic contact within 50 cm of the soil
surface;
b. Have an Ap horizon that has a color value, moist, of 4 or
more or has a value, dry, of 6 or more when crushed and
smoothed (smoothed with a knife to eliminate shadows), or

the A horizon is less than 15 cm thick if its color value, moist, is lower than 3.5; and
c. Do not have an albic horizon at the surface or immediately under an A or Ap horizon that, in turn, is underlain by another horizon that has a color value more than one unit darker or that has chroma of 6 or more.
Humaqueptic Psammaquents are like Typic Psammaquents except for *b*, and their base saturation (by NH_4OAc) is <50 percent in more than half the thickness of the subhorizons within the upper 1 m.
Lithic Psammaquents are like Typic Psammaquents except for *a*.
Mollic Psammaquents are like Typic Psammaquents except for *b*, and their base saturation (by NH_4OAc) is 50 percent or more in more than half the thickness of the subhorizons within the upper 1 m.
Spodic Psammaquents are like Typic Psammaquents except for *c*.

Sulfaquents

Definition of Typic Sulfaquents

Typic Sulfaquents are the Sulfaquents that
a. Have sulfidic materials within 50 cm of the mineral soil surface if the n value is >1 or within 30 cm if the n value is <1.

Tropaquents

The definition of the great group of Tropaquents that follows cannot be tested in the United States and is provisional.

Tropaquents are the Aquents that
1. Have an isomesic or warmer iso temperature regime;
2. Have an n value of 0.7 or less or have <8 percent clay in some subhorizon between 20 and 50 cm; and
3. Have an organic-carbon content[6] that decreases regularly with depth below 25 cm and reaches a level of 0.2 percent or less within a depth of 1.25 m.

Subgroups have not been defined.

Arents

The Arents form a unique suborder in that not only are no great groups recognized, but also there is no typic subgroup. Subgroups of Arents are intergrades to suborders or great groups of Spodosols, Alfisols, or other orders, according to the nature of the fragments that can be identified and according to the soil moisture and soil temperature regimes of the suborders. Thus, if fragments of an argillic horizon can be identified by their clay content and clay skins, the soil might be called an Udalfic Arent if the moisture regime is udic or a Xeralfic Arent if the moisture regime is xeric. At

---

[6]The carbon should be of Holocene age. It is not the intent to include fossil carbon from transported fragments of bedrock or from buried Pleistocene deposits. The mean residence time of the carbon should be <11,000 years B.P.

present, few series have been established for Arents in the United States.

Arents are the Entisols that
1. Have fragments of diagnostic horizons that occur more or less without discernible order in the soil below any Ap horizon but within a depth of 1 m; and
2. Are not permanently saturated with water and do not have the characteristics associated with wetness that are defined for Aquents.

Hapludollic Arents have a udic moisture regime and fragments of a mollic epipedon within the upper 1 m of the soil. *Udalfic Arents* have a udic moisture regime and fragments of an argillic horizon that has base saturation (by sum of cations) that is 35 percent or more within the upper 1 m of the soil.

Fluvents

Key to great groups

JDA. Fluvents that have a cryic soil temperature regime.
Cryofluvents, p. 117

JDB. Other Fluvents that have a xeric moisture regime.
Xerofluvents, p. 120

JDC. Other Fluvents that have an ustic moisture regime.
Ustifluvents, p 119

JDD. Other Fluvents that have a torric moisture regime.
Torrifluvents, p. 118

JDE. Other Fluvents that have an isomesic, isothermic, or isohyperthermic temperature regime.
Tropofluvents, p. 119

JDF. Other Fluvents.
Udifluvents, p. 119

Cryofluvents

Distinctions between Typic Cryofluvents and other subgroups

Typic Cryofluvents are the Cryofluvents that
a. Do not have a layer in the upper 75 cm that has texture finer than loamy fine sand, that is as much as 18 cm thick, that has bulk density in the fine-earth fraction (at 33 kPa moisture tension) of 0.95 g per cubic centimeter or less, and that has either (1) a ratio of measured clay to 1500 kPa water (percentages) of 1.25 or less or (2) a ratio of CEC (at pH near 8) to 1500 kPa water of more than 1.5 and more exchange acidity than the sum of bases plus KCl-extractable aluminum;
b. Do not have mottles that have chroma of 2 or less within 50 cm of the soil surface; and
c. Have an Ap horizon that has a color value, moist, of 4 or more or has a color value, dry, of 6 or more when crushed

and smoothed, or the A horizon is <15 cm thick if its color value, moist, is less than 3.5.

Andeptic Cryofluvents are like Typic Cryofluvents except for *a.*

Aquic Cryofluvents are like Typic Cryofluvents except for *b* or for *b* and *c.*

Mollic Cryofluvents are like Typic Cryofluvents except for *c.*

Torrifluvents

Distinctions between Typic Torrifluvents and other subgroups

Typic Torrifluvents are the Torrifluvents that
a. Do not have a horizon within 1 m of the surface that is >15 cm thick that either contains as much as 20 percent durinodes in a nonbrittle matrix or is brittle and has firm consistence when moist;
b. Do not have all three of the following characteristics:
  (1) Cracks at some period in most years that are 1 cm or more wide at a depth of 50 cm, that are at least 30 cm long in some part, and that extend upward to the soil surface or to the base of an Ap horizon;
  (2) A coefficient of linear extensibility (COLE) of 0.05 or more in a horizon or horizons at least 50 cm thick and a potential linear extensibility of 6 cm or more in the upper 1.5 m of the soil or in the whole soil if a lithic or paralithic contact is deeper than 50 cm but shallower than 1.5 m; and
  (3) More than 35 percent clay in horizons that total >50 cm in thickness within the upper 1 m;
c. Do not have an anthropic epipedon; and
d. Are dry in all parts of the moisture control section three-fourths or more of the time (cumulative) that the soil temperature at a depth of 50 cm is 5°C or more.

Anthropic Torrifluvents are like Typic Torrifluvents except for *c.*

Durorthidic Torrifluvents are like Typic Torrifluvents except for *a.*

Durorthidic Xeric Torrifluvents are like Typic Torrifluvents except for *a* and *d.* They have a thermic, mesic, or frigid soil temperature regime and a torric moisture regime that borders on a xeric regime.

Ustertic Torrifluvents are like Typic Torrifluvents except for *b,* with or without *d.* Unless the soils are irrigated, the cracks remain open in most years from 175 to 240 days, cumulative, and are not closed for as many as 60 consecutive days during the 120 days following the winter solstice in 3 or more years out of 10 if the soil temperature regime is thermic, mesic, or frigid.

Ustic Torrifluvents are like Typic Torrifluvents except for *d,* and they have a torric moisture regime that borders on an ustic regime.

Vertic Torrifluvents are like Typic Torrifluvents except for *b,* with or without *c* or *d* or both. Unless the soils are irrigated, the cracks remain open in most years for more than 240 days, cumulative, and are not closed for as many as 60 consecutive days at any season in most years.

Xeric Torrifluvents are like Typic Torrifluvents except for *d,* and they have a thermic, mesic, or frigid soil temperature regime and a torric moisture regime that borders on a xeric regime.

Tropofluvents

Tropofluvents are the Fluvents that
1. Have an isomesic or warmer iso temperature regime; and
2. Have a udic moisture regime.

Only one series of Tropofluvents has been recognized in the United States, and subgroups have not been defined. It is suggested that the definition of the typic subgroup should parallel that of Typic Udifluvents in the great group that is defined next.

Udifluvents

Distinctions between Typic Udifluvents and other subgroups

Typic Udifluvents are the Udifluvents that

a. Do not have a layer in the upper 75 cm that has texture finer than loamy fine sand, that is as much as 18 cm thick, that has a bulk density in the fine-earth fraction (at 33 kPa moisture tension) of 0.95 g per cubic centimeter or less, and that has either (1) a ratio of measured clay to 1500 kPa water (percentages) of 1.25 or less or (2) a ratio of CEC (at pH near 8) to 1500 kPa water of >1.5 and more exchange acidity than the sum of bases plus KCl-extractable aluminum;
b. Do not have mottles within 50 cm of the surface that have chroma of 2 or less or, at a depth between 50 cm and 1 m, do not have any horizons that are saturated with water at some period or that are artificially drained and have chroma less than 1 or hue bluer than 10Y and value, moist, of 4 or more; and
c. Have an Ap horizon that has a color value, moist, of 4 or more or has a color value, dry, of 6 or more when crushed and smoothed, or the A horizon is <15 cm thick if its color value, moist, is less than 3.5.
Aquic Udifluvents are like Typic Udifluvents except for *b* or for *b* and *c*.
Mollic Udifluvents are like Typic Udifluvents except for *c*.

Ustifluvents

Distinctions between Typic Ustifluvents and other subgroups

Typic Ustifluvents are the Ustifluvents that
a. Do not have mottles within 50 cm of the surface that have chroma of 2 or less and do not have, at a depth within 1.5 m of the surface, a horizon that is saturated with water at some period or is artificially drained and that has chroma less than 1 or a hue bluer than 10Y; and
b. Do not have the following combination of characteristics:
    (1) Cracks at some period in most years, when the soil is not irrigated, that are 1 cm or more wide at a depth of 50 cm, that are at least 30 cm long in some part, and that extend upward to the soil surface or to the base of an Ap horizon;
    (2) A coefficient of linear extensibility (COLE) of 0.07 or more in a horizon or horizons at least 50 cm thick and

a potential linear extensibility of 6 cm or more in the upper 1.25 m of the soil or in the whole soil if a lithic or paralithic contact is deeper than 50 cm but shallower than 1.25 m; and

(3) More than 35 percent clay in horizons that total >50 cm in thickness.

c. Have an Ap horizon that has a color value, moist, of 4 or more or has a color value, dry, of 6 or more when crushed and smoothed, or the A horizon is <15 cm thick if its color value, moist, is less than 3.5

Aquic Ustifluvents are like Typic Ustifluvents except for *a* or for *a* and *c*.

Mollic Ustifluvents are like Typic Ustifluvents except for *c*.

Vertic Ustifluvents are like Typic Ustifluvents except for *b* or for *b* and *a*.

Xerofluvents

Distinctions between Typic Xerofluvents and other subgroups

Typic Xerofluvents are the Xerofluvents that
a. Are not saturated with water within 1.5 m of the surface during any period in most years;
b. Do not have a horizon within 1 m of the surface that is >15 cm thick that either contains 20 percent or more durinodes in a nonbrittle matrix or that is brittle and has firm consistence when moist;
c. Do not have the following combination of characteristics:
(1) Cracks at some period in most years that are 1 cm or more wide at a depth of 50 cm, that are at least 30 cm long in some part, and that extend upward to the soil surface or to the base of an Ap horizon;
(2) A coefficient of linear extensibility (COLE) of 0.05 or more in a horizon or horizons at least 50 cm thick and a potential linear extensibility of 6 cm or more in the upper 1.5 m of the soil or in the whole soil if a lithic or paralithic contact is deeper than 50 cm but shallower than 1.5 m; and
(3) More than 35 percent clay in horizons that total >50 cm in thickness; and
d. Have an Ap horizon that has a color value, moist, of 4 or more or has a color value, dry, of 6 or more when crushed and smoothed, or the A horizon is <15 cm thick if its color value, moist, is less than 3.5.

Aquic Xerofluvents are like Typic Xerofluvents except for *a* or for *a* and *d*.

Aquic Durorthidic Xerofluvents are like Typic Xerofluvents except for *a* and *b* with or without d.

Mollic Xerofluvents are like Typic Xerofluvents except for *d*.

Vertic Xerofluvents are like Typic Xerofluvents except for *c* or for *c* and *a*.

Orthents

Key to great groups

JEA. Orthents that have a cryic or pergelic temperature regime.

Cryorthents, p. 121

JEB. Other Orthents that have a torric moisture regime or that have a conductivity of the saturation extract that is 2 dS m^{-1} or greater at 25°C in some part above whichever of the following depths is the least: a lithic or paralithic contact or 1.25 m if particle-size class[7] is sandy, 90 cm if loamy, and 75 cm if clayey.

Torriorthents, p. 112

JEC. Other Orthents that have a xeric moisture regime.

Xerorthents, p. 124

JED. Other Orthents that have a udic moisture regime and mean summer and mean winter soil temperatures at a depth of 50 cm that differ by <5°C.

Troporthents, p. 123

JEE. Other Orthents that have a udic moisture regime.

Udorthents, p. 123

JEF. Other Orthents.

Ustorthents, p. 124

Cryorthents

Distinctions between Typic Cryorthents and other subgroups

Typic Cryorthents are the Cryorthents that
a. Do not have a layer in the upper 75 cm that has texture finer than loamy fine sand, that is as much as 18 cm thick, that has bulk density in the fine-earth fraction (at 33 kPa water tension) of 0.95 g per cubic centimeter or less and that has either
    (1) a ratio of measured clay to 1500 kPa water (percentages) of 1.25 or less or
    (2) a ratio of CEC (at pH near 8) to 1500 kPa water >1.5 and more exchange acidity than the sum of bases plus KCl-extractable aluminum;
b. Do not have mottles that have chroma of 2 or less within 50 cm of the soil surface;
c. Do not have a lithic contact within 50 cm of the soil surface;
d. Have a mean annual soil temperature that is higher than 0°C; and
e. Do not have lamellae within 1.5 m of the soil surface that meet all requirements for an argillic horizon except thickness[8].
Alfic Cryorthents are like Typic Cryorthents except for *e.*
Alfic Andeptic Cryorthents are like Typic Cryorthents except for *a,* and *e.*
Andeptic Cryorthents are like Typic Cryorthents except for *a.*
Aquic Cryorthents are like Typic Cryorthents except for *b.*
Lithic Cryorthents are like Typic Cryorthents except for *c,* with or without *d.*

---

[7]The weighted average particle-size class between a depth of 1 m or a lithic or paralithic contact, whichever is shallower.

[8]The clay content cannot be estimated with precision in lamellae that are very thin. The lamellae in soils of alfic subgroups generally are about 0.5 to 1 cm thick, but their total thickness is less than the 15 cm required for an argillic horizon.

Pergelic Cryorthents are like Typic Cryorthents except for *d*, with or without *a* or *b*, or both.

Torriorthents

Distinctions between Typic Torriorthents and other subgroups

Typic Torriorthents are the Torriorthents that
a. Do not have a horizon within 1 m of the surface that is >15 cm thick that either contains 20 percent or more durinodes in a nonbrittle matrix or is brittle and has firm consistence when moist;
b. Do not have a lithic contact within 50 cm of the surface;
c. Do not have the following combination of characteristics:
   (1) Cracks at some period in most years that are 1 cm or more wide at a depth of 50 cm, that are at least 30 cm long in some part, and that extend upward to the soil surface or the base of an Ap horizon;
   (2) A coefficient of linear extensibility (COLE) of 0.05 or more in a horizon or horizons at least 50 cm thick and a potential linear extensibility of 6 cm or more in the upper 1.5 m of the soil or in the whole soil if a lithic or paralithic contact is deeper than 50 cm but shallower than 1.5 m; and
   (3) More than 35 percent clay in horizons that total >50 cm in thickness;
d. Are dry in all parts of the moisture control section three-fourths or more of the time (cumulative) that the soil temperature at a depth of 50 cm is 5°C or higher; and
e. Are not saturated with water within 1.5 m of the surface at any time of year in most years.
Aquic Torriorthents are like Typic Torriorthents except for *e* or for *d* and *e*.
Aquic Durorthidic Torriorthents are like Typic Torriorthents except for *a* and *e*, or for *a*, *d*, and *e*.
Durorthidic Torriorthents are like Typic Torriorthents except for *a*.
Durorthidic Xeric Torriorthents are like Typic Torriorthents except for *a* and *d*. They have a thermic, mesic, or frigid soil temperature regime and have a torric moisture regime that borders on a xeric regime.
Lithic Torriorthents are like Typic Torriorthents except for *b*.
Lithic Ustic Torriorthents are like Typic Torriorthents except for *b* and *d*. They either (1) have a hyperthermic or an iso soil temperature regime or (2) have a thermic or mesic soil temperature regime and have a torric moisture regime that borders on an ustic regime.
Lithic Xeric Torriorthents are like Typic Torriorthents except for *b* and *d*. They have a thermic, mesic, or frigid soil temperature regime and an aridic moisture regime that borders on a xeric regime.
Ustertic Torriorthents are like Typic Torriorthents except for *c*, with or without *d* or *e*, or both. Unless irrigated, they have cracks that remain open from 175 to 240 days, cumulative, and are not closed for as many as 60 consecutive days during the 120 days following the winter solstice in 3 or more years out of 10 if the soil temperature regime is thermic, mesic, or frigid.

Ustic Torriorthents are like Typic Torriorthents except for *d*. They either (1) have a hyperthermic or an iso soil temperature regime or (2) have a thermic, mesic, or a frigid soil temperature regime and have an aridic moisture regime that borders on an ustic regime.

Vertic Torriorthents are like Typic Torriorthents except for *c* or for *c* and *d*. Unless the soil is irrigated, the cracks remain open in most years for more than 240 days, cumulative, and are not closed in most years for as many as 60 consecutive days at any season.

Xerertic Torriorthents are like Typic Torriorthents except for *c*, with or without *d* or *e*, or both. They have a thermic, mesic, or frigid soil temperature regime and have cracks that are closed for 60 consecutive days or more during the 120 days following the winter solstice in more than 7 years out of 10.

Xeric Torriorthents are like Typic Torriorthents except for *d*. They have a thermic, mesic, or frigid soil temperature regime and a torric moisture regime that borders on a xeric regime.

Troporthents

Distinctions between Typic Troporthents and other subgroups

Typic Troporthents are the Troporthents that
a. Do not have a layer in the upper 75 cm that has texture finer than loamy fine sand, that is as much as 18 cm thick, that has a bulk density in the fine-earth fraction (at 33 kPa moisture tension) of 0.95 g per cubic centimeter or less, and that has either
   (1) a ratio of measured clay to 1500 kPa water (percentages) of 1.25 or less or
   (2) a ratio of CEC (at pH near 8) to 1500 kPa water of more than 1.5 and more exchange acidity than the sum of bases plus KCl-extractable aluminum; and
b. Do not have a lithic contact within 50 cm of the soil surface.

Andeptic Troporthents are like Typic Troporthents except for *a*.

Lithic Troporthents are like Typic Troporthents except for *b*.

Udorthents

Distinctions between Typic Udorthents and other subgroups

Typic Udorthents are the Udorthents that
a. Do not have a layer in the upper 75 cm that has texture finer than loamy fine sand, that is as much as 18 cm thick, that has a bulk density in the fine-earth fraction (at 33 kPa moisture tension) of 0.95 g per cubic centimeter or less, and that has either
   (1) a ratio of measured clay to 1500 kPa water (percentages) of 1.25 or less, or
   (2) a ratio of CEC (at pH near 8) to 1500 kPa water >1.5 and more exchange acidity than the sum of bases plus KCl-extractable aluminum.
b. Are not saturated with water for as long as 1 month within 1.5 m of the surface;

c. Do not have a lithic contact within 50 cm of the surface; and
d. Have <50 percent by volume of wormholes, wormcasts, and filled animal burrows between the bottom of the Ap horizon or a depth of 25 cm, whichever is deeper, and either a depth of 1 m or a lithic or paralithic contact if one is present above a depth of 1 m.

Andeptic Udorthents are like Typic Udorthents except for *a.*
Aquic Udorthents are like Typic Udorthents except for *b.*
Lithic Udorthents are like Typic Udorthents except for *c* or for *c* and *d.*
Vermic Udorthents are like Typic Udorthents except for *d.*

Ustorthents

Distinctions between Typic Ustorthents and other subgroups

Typic Ustorthents are the Ustorthents that
a. Are not saturated with water within 1.5 m of the surface for as long as 1 month in most years;
b. Do not have a horizon within 1 m of the surface that is >15 cm thick that either contains 20 percent or more durinodes in a nonbrittle matrix or is brittle and has firm consistence when moist;
c. Do not have a lithic contact within 50 cm of the surface;
d. Have <50 percent (by volume) wormholes, wormcasts, and filled animal burrows between the bottom of the Ap horizon or a depth of 25 cm, whichever is deeper, and a depth of 1 m or a lithic or paralithic or petroferric contact, whichever is shallower; and
e. Do not have the following combination of characteristics:
  (1) Cracks at some period in most years that are 1 cm or more wide at a depth of 50 cm, that are at least 30 cm long in some part, and that extend upward to the soil surface or to the base of an Ap horizon; and
  (2) A coefficient of linear extensibility (COLE) of 0.07 or more in a horizon or horizons at least 50 cm thick and a potential linear extensibility of 6 cm or more in the upper 1.25 m of the soil or in the whole soil if a lithic or paralithic contact is deeper than 50 cm but shallower than 1.25 m; and
  (3) More than 35 percent clay in horizons that total >50 cm in thickness.

Aquic Ustorthents are like Typic Ustorthents except for *a.*
Lithic Ustorthents are like Typic Ustorthents except for *c.*
Vermic Ustorthents are like Typic Ustorthents except for *d.*
Vertic Ustorthents are like Typic Ustorthents except for *e.*

Xerorthents

Distinctions between Typic Xerorthents and other subgroups

Typic Xerorthents are the Xerorthents that
a. Are not saturated with water within 1.5 m of the surface at any time of year in most years;
b. Do not have a horizon within 1 m of the surface that is >15 cm thick, that either contains 20 percent or more durinodes in a nonbrittle matrix or is brittle and has firm consistence when moist;
c. Do not have a lithic contact within 50 cm of the soil surface; and

d. Have base saturation (by NH_4OAc) of 60 percent or more in some part of the soil between depths of 25 cm and 75 cm below the soil surface.

Aquic Xerorthents are like Typic Xerorthents except for *a*.
Aquic Durorthidic Xerorthents are like Typic Xerorthents except for *a* and *b*.
Durorthidic Xerorthents are like Typic Xerorthents except for *b*.
Dystric Xerorthents are like Typic Xerorthents except for *d*.
Lithic Xerorthents are like Typic Xerorthents except for *c* or for *c* and *d*.

Psamments

Key to great groups

JCA. Psamments that have a cryic or pergelic soil temperature regime.

Cryopsamments, p. 125

JCB. Other Psamments that have a torric moisture regime.

Torripsamments, p. 127

JCC. Other Psamments that have, in the particle size control section, more than 90 percent silica minerals (quartz, chalcedony or opal) or other extremely durable minerals in the 0.02 to 2mm fraction that are resistant to weathering.

Quartzipsamments, p. 126

JCD. Other Psamments that have a udic moisture regime and mean summer and mean winter soil temperatures at a depth of 50 cm that differ by 5°C or more.

Udipsamments, p. 127

JCE. Other Psamments that have a udic moisture regime.

Tropopsamments, p. 127

JCF. Other Psamments that have a xeric moisture regime.

Xeropsamments, p. 129

JCG. Other Psamments.

Ustipsamments, p. 128

Cryopsamments

Distinctions between Typic Cryopsamments and other subgroups

Typic Cryopsamments are the Cryopsamments that
a. Do not have lamellae within 1.5 m of the soil surface that meet all requirements for an argillic horizon except thickness;
b. Do not have mottles that have chroma of 2 or less within 50 cm of the soil surface;
c. Have a mean annual soil temperature that is higher than 0°C;
d. Do not have a lithic contact within 50 cm of the soil surface; and
e. Do not have an albic horizon that is 5 cm or more thick and underlain by a horizon that has a color value one unit

or more darker and that meets all requirements of a spodic horizon except the index of accumulation.

Alfic Cryopsamments are like Typic Cryopsamments except for *a*.

Aquic Cryopsamments are like Typic Cryopsamments except for *b*.

Lithic Cryopsamments are like Typic Cryopsamments except for *d* or for *c* and *d*.

Pergelic Cryopsamments are like Typic Cryopsamments except for *c*, with or without *b*.

Spodic Cryopsamments are like Typic Cryopsamments except for *e*.

Quartzipsamments

Distinctions between Typic Quartzipsamments and other subgroups

Typic Quartzipsamments are the Quartzipsamments that
a. Do not have mottles above a depth of 1 m that have chroma of 2 or less or, if the color is due to uncoated sand grains, do not have the water table within 1 m of the soil surface for as many as 60 days, cumulative, in most years;
b. Do not have an albic horizon[9] at the surface or immediately under an A or an Ap horizon that is underlain by another horizon that has a color value more than one unit darker or chroma of 6 or more;
c. Do not have a lithic contact within 50 cm of the soil surface;
d. Have a clay fraction that has a higher CEC than that of the clay of an oxic horizon or >25 percent of the surfaces of sand grains are uncoated;
e. Have <5 percent plinthite in all horizons above a depth of 1 m; and
f. Have a udic moisture regime.

Aquic Quartzipsamments are like Typic Quartzipsamments except for *a*.

Haplaquodic Quartzipsamments are like Typic Quartzipsamments except for *a* and *b*. They have a ground-water table that is within 1 m of the soil surface for 6 months or more in most years or are artificially drained, and they have a difference of 5°C or more between the mean summer and mean winter soil temperatures at a depth of 50 cm.

Lithic Quartzipsamments are like Typic Quartzipsamments except for *c*.

Orthoxic Quartzipsamments are like Typic Quartzipsamments except for *d*, with or without *e*, and they have enough clay to coat at least 75 percent of the surfaces of the sand grains.

Spodic Quartzipsamments are like Typic Quartzipsamments except for *b*.

Tropaquodic Quartzipsamments are like Typic Quartzipsamments except for *a* and *b*, and they have a difference of <5°C between the mean winter and mean summer soil temperatures at a depth of 50 cm.

Ustic Quartzipsamments are like Typic Quartzipsamments except for *f*.

Ustoxic Quartzipsamments are like Typic Quartzipsamments except for *d* and *f*, with or without *e*. They have an ustic

---

[9]The albic horizon must be thick enough to be preserved after the soil to a depth of 18 cm is mixed.

moisture regime and have enough clay to coat at least 75 percent of the surfaces of the sand grains.

Torripsamments

Distinctions between Typic Torripsamments and other subgroups

Typic Torripsamments are the Torripsamments that
a. Do not have a lithic contact within 50 cm of the soil surface;
b. Do not have a horizon within 1 m of the surface that is >15 cm thick and either contains 20 percent or more durinodes in a nonbrittle matrix or is brittle and has firm consistence when moist; and
c. Are dry in all parts of the moisture control section three-fourths or more of the time (cumulative) that the soil temperature at a depth of 50 cm is 5°C or higher.
Durorthidic Xeric Torripsamments are like Typic Torripsamments except for *b* and *c*, and they have a thermic, mesic, or frigid soil temperature regime and a torric moisture regime that borders on xeric.
Lithic Torripsamments are like Typic Torripsamments except for *a* or for *a* and *c*.
Ustic Torripsamments are like Typic Torripsamments except for *c*, and they have a torric moisture regime that borders on an ustic regime.
Xeric Torripsamments are like Typic Torripsamments except for *c*, and they have a thermic, mesic, or frigid soil temperature regime and a torric moisture regime that borders on an xeric regime.

Tropopsamments

Distinctions between Typic Tropopsamments and other subgroups

Typic Tropopsamments are the Psamments that
a. Do not have a lithic contact within 50 cm of the soil surface; and
b. Do not have mottles above a depth of 1 m that have chroma of 2 or less, or if the color is due to uncoated sand grains, have a ground-water table within 1 m of the soil surface for less than 60 cumulative days in most years.
Aquic Tropopsamments are like Typic Tropopsamments except for *b*.
Lithic Tropopsamments are like Typic Tropopsamments except for *a*.

Udipsamments

Distinctions between Typic Udipsamments and other subgroups

Typic Udipsamments are the Udipsamments that
a. Do not have lamellae within 1.5 m of the soil surface that meet all requirements for an argillic horizon except thickness[10];

---

[10]The clay content cannot be estimated with precision in lamellae that are very thin. The lamellae in soils of alfic subgroups generally are about 0.5 to 1

b. Do not have mottles that have chroma of 2 or less above a depth of 1 m;

c. Do not have a lithic contact within a depth of 50 cm;

d. Do not have an albic horizon that is thick enough to be preserved after the soil has been mixed to a depth of 18 cm and is underlain by a horizon that has a color value one unit or more darker and that meets all requirements for a spodic horizon except the index of accumulation; and

e. Do not have a surface horizon between 25 and 50 cm thick that meets all requirements for a plaggen epipedon except thickness.

Alfic Udipsamments are like Typic Udipsamments except for *a* and have base saturation of 35 percent or more in some horizon <1.25 m below the uppermost lamella or have a frigid temperature regime.

Aquic Udipsamments are like Typic Udipsamments except for *b* with or without *a* or *d* or both.

Lithic Udipsamments are like Typic Udipsamments except for *c*.

Plaggeptic Udipsamments are like Typic Udipsamments except for *e*.

Spodic Udipsamments are like Typic Udipsamments except for *d*.

Ultic Udipsamments are like Typic Udipsamments except for *a*, have base saturation of <35 percent in some horizon <1.25 m below the uppermost lamella, and have a mesic or warmer temperature regime.

Ustipsamments

Distinctions between Typic Ustipsamments and other subgroups

Typic Ustipsamments are the Ustipsamments that
a. Do not have lamellae within 1.5 m of the soil surface that meet all requirements for an argillic horizon except thickness[11];

b. Do not have distinct or prominent mottles above a depth of 1 m and are not saturated with water within 1 m of the surface during any time of year in most years; and

c. Do not have a lithic contact within 50 cm of the surface.

Alfic Ustipsamments are like Typic Ustipsamments except for *a*, have base saturation of 35 percent of more in some horizon <1.25 m below the uppermost lamella, and have a color value, moist, of 4 or more within 25 cm of the surface.

Aquic Ustipsamments are like Typic Ustipsamments except for *b*.

Lithic Ustipsamments are like Typic Ustipsamments except for *c*.

---

cm thick, but their total thickness is less than the 15 cm required for an argillic horizon.

[11] The clay content cannot be estimated with precision in lamellae that are very thin. The lamellae in soils of alfic subgroups generally are about 0.5 to 1 cm thick, but their total thickness is less than the 15 cm required for an argillic horizon.

Xeropsamments

Distinctions between Typic Xeropsamments and other subgroups

Typic Xeropsamments are the Xeropsamments that
a. Do not have lamellae within 1.5 m of the soil surface that
meet all requirements for an argillic horizon except thickness[12];
b. Do not have distinct or prominent mottles above a depth
of 1 m and are not saturated with water within 1 m of the
surface during any time of year in most years;
c. Do not have a horizon within 1 m of the surface that is
>15 cm thick and that either contains 20 percent or more
durinodes in a nonbrittle matrix or is brittle and has firm
consistence when moist;
d. Do not have a lithic contact within 50 cm of the soil
surface; and
e. Have base saturation (by NH_4OAc) of 60 percent or more
in some part of the soil between depths of 25 cm and 75 cm
below the soil surface.
Alfic Xeropsamments are like Typic Xeropsamments except
for *a* and have base saturation of 35 percent or more in
some horizon <1.25 m below the uppermost lamella.
Aquic Xeropsamments are like Typic Xeropsamments except
for *b*.
Aquic Durorthidic Xeropsamments are like Typic Xeropsamments except for *b* and *c*.
Dystric Xeropsamments are like Typic Xeropsamments except for *e*.
Lithic Xeropsamments are like Typic Xeropsamments except
for *d* or for *d* and *e*.

---

[12]The clay content cannot be estimated with precision in lamellae that are
very thin. The lamellae in soils of alfic subgroups generally are about 0.5 to 1
cm thick, but their total thickness is less than the 15 cm required for an
argillic horizon.

Chapter 8

Histosols

Key To Suborders

AA. Histosols that

1. Are never saturated with water for more than a few days following heavy rains and

    a. Have a lithic or paralithic contact <1 m from the surface or have fragmental materials in which the interstices are filled or partly filled with organic materials in half or more of each pedon, or both; and

    b. Less than three-fourths of the thickness of organic materials consists of Sphagnum fibers.

Folists, p. 136

AB. Other Histosols that

1. Are dominantly[1] fibric in the subsurface tier if that tier is wholly organic except for a thin mineral layer or layers, or the organic parts of the surface and subsurface tiers are dominantly fibric if a continuous mineral layer 40 cm or more thick begins within the depth limit of the subsurface tier; or

2. Have a surface mantle that has three-fourths or more of its volume consisting of fibers derived from *Sphagnum* and that rests on a lithic or paralithic contact, fragmental materials, or mineral soil, or on frozen[2] materials within the limits in depth of the surface or subsurface tier; and

3. Do not have a sulfuric horizon whose upper boundary is within 50 cm of the surface and do not have sulfidic materials within 1 m of the surface.

Fibrists, p. 132

AC. Other Histosols that

1. Are dominantly hemic in the subsurface tier if that tier is wholly organic except for a thin mineral layer or layers; or are dominantly hemic in the organic part of the surface and subsurface tiers if a continuous mineral layer 40 cm or more thick begins within the depth limits of the subsurface tier; or

2. Have a sulfuric horizon whose upper boundary is within 50 cm of the surface or have sulfidic materials within 1 m of the surface.

Hemists, p. 137

AD. Other Histosols.

Saprists, p. 140

---

[1] Dominant, in this context, means the most abundant. If only two kinds of organic materials are present, the fibric materials occupy half or more of the volume. If there are both hemic and sapric materials as well as fibric, the fibric materials may occupy less than half of the volume but have more volume than either the hemic or sapric materials.

[2] Frozen 2 months after the summer solstice.

Fibrists

Key to great groups

ABA. Fibrists that have a surface mantle that is three-fourths or more fibric *Sphagnum* spp. and that either is 90 cm or more thick, or extends 10 cm or more below permafrost, or rests on a lithic or paralithic contact, fragmental materials, or mineral soil materials.

Sphagnofibrists, p. 134

ABB. Other Fibrists that are frozen in most years in some layer within the control section about 2 months after the summer solstice or that are never frozen in most years below a depth of 5 cm but have a mean annual soil temperature that is lower than 8°C.

Cryofibrists, p. 133

ABC. Other Fibrists that have a mean annual soil temperature lower than 8°C.

Borofibrists, p. 132

ABD. Other Fibrists that have a difference of <5°C between mean summer and mean winter soil temperatures at a depth of 30 cm.

Tropofibrists, p. 135

ABE. Other Fibrists that do not have a horizon 2 cm or more thick that is half or more humilluvic materials.

Medifibrists, p. 134

ABF. Other Fibrists.

Luvifibrists, p. 133

Borofibrists

Distinctions between Typic Borofibrists and other subgroups

Typic Borofibrists are the Borofibrists that
a. Have
    (1) Less than 25 cm of the subsurface and bottom tiers occupied by hemic materials; and
    (2) Less than 12.5 cm of the subsurface and bottom tiers occupied by sapric materials;
b. Have less than three-fourths of the fibers (by volume) derived from *Sphagnum* in the surface tier or more of the control section;
c. Do not have limnic layer(s) within the control section 5 cm or more thick;
d. Do not have a lithic contact within the control section;
e. Do not have a mineral layer between 5 and 30 cm thick within organic materials or do not have two or more thin continuous mineral layers in the control section below the surface tier;
f. Do not have a mineral layer 30 cm or more thick whose upper boundary is below the surface tier in the control section; and
g. Do not have a layer of water within the control section beneath the surface tier.
Fluvaquentic Borofibrists are like Typic Borofibrists except for *e* with or without a.

Hemic Borofibrists are like Typic Borofibrists except for
a(1).
Hemic Terric Borofibrists are like Typic Borofibrists except
for *a(1)* and *f*, with or without *c* or *e*, or both.
Hydric Borofibrists are like Typic Borofibrists except for *g*.
Limnic Borofibrists are like Typic Borofibrists except for *c*,
with or without *a* or *e*, or both.
Lithic Borofibrists are like Typic Borofibrists except for *d*,
with or without all or any of *a*, *b*, *c*, *e*, or *f*.
Sapric Borofibrists are like Typic Borofibrists except for
a(2).
Sapric Terric Borofibrists are like Typic Borofibrists except
for *a(2)* and *f*, with or without *c* or *e*, or both.
Sphagnic Borofibrists are like Typic Borofibrists except for
b.
Sphagnic Terric Borofibrists are like Typic Borofibrists except for *b* and *f*, with or without *c* or *e*, or both.
Terric Borofibrists are like Typic Borofibrists except for *f*,
with or without *c* or *e*, or both.

Cryofibrists

Distinctions between Typic Cryofibrists and other subgroups

Typic Cryofibrists are the Cryofibrists that
a. Have less than three-fourths of their fiber volume derived from *Sphagnum* spp. in the surface tier or more of the
control section;
b. Have a mean annual soil temperature higher than 0°C;
c. Do not have a lithic contact within the control section;
d. Do not have a mineral layer between 5 and 30 cm thick
with organic materials or do not have two or more thin,
continuous mineral layers in the control section below the
surface tier;
e. Do not have a mineral layer 30 cm or more thick whose
upper boundary is below the surface tier in the control section; and
f. Have organic materials that are continuous laterally
throughout each pedon in at least the surface tier.
Fluvaquentic Cryofibrists are like Typic Cryofibrists except
for *d*.
Lithic Cryofibrists are like Typic Cryofibrists except for *c*,
with or without *a* or both *a* and *b*.
Pergelic Cryofibrists are like Typic Cryofibrists except for
b, with or without any or all or *a*, *d*, or *e*.
Sphagnic Cryofibrists are like Typic Cryofibrists except for
a.
Terric Cryofibrists are like Typic Cryofibrists except for *e*
or for *d* and *e*.

Luvifibrists

Luvifibrists are not known to occur in the United States, but
the great group is provided tentatively for use in other
countries if needed. These soils are the Fibrists that have a
horizon within the control section that is 2 cm or more thick
and is half or more humilluvic materials. Because these soils
cannot be studied in the United States, a precise definition
is not attempted here. It should be noted, however, that the
soils normally are acid and that they have been cultivated
for a long time. been cultivated for a long time.

Medifibrists

Distinctions between Typic Medifibrists and other subgroups

Typic Medifibrists are the Medifibrists that
a. Have
   (1) Less than 25 cm of the subsurface and bottom tiers occupied by hemic materials; and
   (2) Less than 12.5 cm of the subsurface and bottom tiers occupied by sapric materials;
b. Have less than three-fourths of the fiber volume in the surface tier or more of the control section derived from *Sphagnum*;
c. Do not have limnic layer(s) that are 5 cm or more thick within the control section;
d. Do not have a lithic contact within the control section;
e. Do not have a mineral layer between 5 cm and 30 cm thick within organic materials or do not have two or more thin continuous mineral layers in the control section below the surface tier;
f. Do not have a mineral layer 30 cm or more thick whose upper boundary in the control section is below the surface tier; and
g. Do not have a layer of water within the control section beneath the surface tier.
Fluvaquentic Medifibrists are like Typic Medifibrists except for *e*, with or without *a*.
Hemic Medifibrists are like Typic Medifibrists except for *a(1)*.
Hemic Terric Medifibrists are like Typic Medifibrists except for *a(1)* and *f*, with or without *c* or *e*, or both.
Hydric Medifibrists are like Typic Medifibrists except for *g*.
Limnic Medifibrists are like Typic Medifibrists except for *c*, with or without *e* or *a*, or both.
Lithic Medifibrists are like Typic Medifibrists except for *d*, with or without all or any of *a*, *b*, *c*, *e*, or *f*.
Sapric Medifibrists are like Typic Medifibrists except for *a(2)*.
Sapric Terric Medifibrists are like Typic Medifibrists except for *a(2)*, and *f*, with or without *c* or *e*, or both.
Sphagnic Medifibrists are like Typic Medifibrists except for *b*.
Sphagnic Terric Medifibrists are like Typic Medifibrists except for *b* and *f*, with or without *c* or *e*, or both.
Terric Medifibrists are like Typic Medifibrists except for *f*, with or without *c* or *e*, or both.

Sphagnofibrists

Distinctions between Typic Sphagnofibrists and other subgroups

Typic Sphagnofibrists are the Sphagnofibrists that
a. Do not have a mineral layer between 5 and 30 cm thick within organic materials or do not have two or more thin continuous mineral layers in the control section below the surface tier;
b. Have
   (1) Less than 25 cm of the subsurface and bottom tiers occupied by hemic materials; and

(2) Less than 12.5 cm of the subsurface and bottom tiers occupied by sapric materials;
c. Do not have a layer of water within the control section beneath the surface tier;
d. Do not have limnic layer(s) that are 5 cm or more thick within the control section.
e. Do not have a lithic contact within the control section;
f. Do not have a mineral layer 30 cm or more thick that has its upper boundary in the control section below the surface tier;
g. Are never frozen within the control section about 2 months after the summer solstice or are never frozen below a depth of 5 cm in most years;
h. Have a mean annual soil temperature higher than 0°C.
Cryic Sphagnofibrists are like Typic Sphagnofibrists except for *g* and have a mean annual soil temperature lower than 8°C but higher than 0°C.
Fluvaquentic Sphagnofibrists are like Typic Sphagnofibrists except for *a*, with or without *b*.
Hemic·Sphagnofibrists are like Typic Sphagnofibrists except for *b(1)*.
Hydric Sphagnofibrists are like Typic Sphagnofibrists except for *c*.
Limnic Sphagnofibrists are like Typic Sphagnofibrists except for *d*, with or without *a* or *b*, or both.
Lithic Sphagnofibrists are like Typic Sphagnofibrists except for *e*, with or without all or any of *a*, *b*, *d*, *f*, or *h*.
Pergelic Sphagnofibrists are like Typic Sphagnofibrists except for *h* and *g*, with or without all or any of *a*, *b*, *d*, *e*, or *f*.
Sapric Sphagnofibrists are like Typic Sphagnofibrists except for *b(2)*.
Terric Sphagnofibrists are like Typic Sphagnofibrists except for *f*, with or without *a* or *d*, or both.

Tropofibrists

Distinctions between Typic Tropofibrists and other subgroups

Typic Tropofibrists are the Tropofibrists that
a. Have
    (1) Less than 25 cm of the thickness of the subsurface and bottom tiers occupied by hemic materials, and
    (2) Less than 12.5 cm of the thickness of the subsurface and bottom tiers occupied by sapric materials;
b. Have less than three-fourths of their fibers, by volume, derived from *Sphagnum* in the surface tier or in more of the control section;
c. Do not have limnic layer(s) that are 5 cm or more thick within the control section;
d. Do not have a lithic contact within the control section;
e. Do not have a mineral layer between 5 and 30 cm thick within organic materials or do not have two or more thin, continuous mineral layers in the control section below the surface tier;
f. Do not have a mineral layer 30 cm or more thick that has its upper boundary in the control section below the surface tier; and
g. Do not have a layer of water within the control section beneath the surface tier.

Fluvaquentic Tropofibrists are like Typic Tropofibrists except for *e*, with or without *a*.
Hemic Tropofibrists are like Typic Tropofibrists except for *a(1)*.
Hemic Terric Tropofibrists are like Typic Tropofibrists except for *a(1)* and *f*, with or without *c* or *e*, or both.
Hydric Tropofibrists are like Typic Tropofibrists except for *g*.
Limnic Tropofibrists are like Typic Tropofibrists except for *c*, with or without *a* or *e*, or both.
Lithic Tropofibrists are like Typic Tropofibrists except for *d*, with or without all or any of *a*, *b*, *c*, *e*, or *f*.
Sapric Tropofibrists are like Typic Tropofibrists except for *a(2)*.
Sapric Terric Tropofibrists are like Typic Tropofibrists except for *a(2)* and *f*, with or without *c* or *e*, or both.
Terric Tropofibrists are like Typic Tropofibrists except for *f*, with or without *c* or *e*, or both.

Folists

Key to great groups

AAA. Folists that have a cryic or colder temperature regime.
Cryofolists, p. 136

AAB. Other Folists that have an isomesic or warmer temperature regime.
Tropofolists, p. 137

AAC. Other Folists that have a frigid temperature regime.
Borofolists, p. 136

Borofolists

Distinctions between Typic Borofolists and other subgroups
Typic Borofolists are the Borofolists that
a. Have fragmental materials with interstices filled with organic materials in half or more of each pedon; and
b. Do not have a lithic contact within 1 m of the surface.
Lithic Borofolists are like the Typic Borofolists except for *b* or for *a* and *b*.

Cryofolists

Distinctions between Typic Cryofolists and other subgroups

Typic Cryofolists are the Cryofolists that
a. Have fragmental materials in which the interstices are filled or partly filled with organic materials in half or more of each pedon; and
b. Do not have a lithic contact within 1 m of the surface.
Lithic Cryofolists are like Typic Cryofolists except for *b* or for *a* and *b*.

Tropofolists

Distinctions between Typic Tropofolists and other subgroups

Typic Tropofolists are the Tropofolists that
a. Have fragmental materials in which the interstices are filled or partly filled with organic materials in half or more of each pedon; and
b. Do not have a lithic contact within 1 m of the surface.
Lithic Tropofolists are like Typic Tropofolists except for *b* or for *a* and *b*.

Hemists

Key to great groups

ACA. Hemists that have a sulfuric horizon that has its upper boundary within 50 cm of the surface.
Sulfohemists, p. 140

ACB. Other Hemists that have sulfidic materials within 1 m of the surface.
Sulfihemists, p. 139

ACC. Other Hemists that have a horizon 2 cm or more thick in which half or more of the volume is humilluvic materials.
Luvihemists, p. 138

ACD. Hemists that are frozen in some layers within the control section about 2 months after the summer solstice in most years or that are never frozen below a depth of 5 cm in most years but have a mean annual soil temperature lower than 8°C.
Cryohemists, p. 138

ACE. Other Hemists that have a mean annual soil temperature lower than 8°C.
Borohemists, p. 137

ACF. Other Hemists that have a difference of <5°C between mean summer and mean winter soil temperatures at a depth of 30 cm.
Tropohemists, p. 140

ACG. Other Hemists.
Medihemists, p. 139

Borohemists

Distinctions between Typic Borohemists and other subgroups

Typic Borohemists are the Borohemists that
a. Do not have a mineral layer between 5 and 30 cm thick within organic materials or do not have two or more thin, continuous mineral layers in the control section below the surface tier;
b. Have
    (1) Less than 25 cm of the subsurface and bottom tiers consisting of fibric materials; and

(2) Less than 25 cm of the subsurface and bottom tiers consisting of sapric materials;

c. Do not have a mineral layer 30 cm or more thick that has its upper boundary in the control section below the surface tier;

d. Do not have a layer of water within the control section beneath the surface tier;

e. Do not have limnic layer(s) 5 cm or more thick within the control section; and

f. Do not have a lithic contact within the control section.

Fibric Borohemists are like Typic Borohemists except for *b(1)*.

Fibric Terric Borohemists are like Typic Borohemists except for *b(1)* and *c*, with or without *a* or *e*, or both.

Fluvaquentic Borohemists are like Typic Borohemists except for *a*, with or without *b*.

Hydric Borohemists are like Typic Borohemists except for *d*.

Limnic Borohemists are like Typic Borohemists except for *e*, with or without *a* or *b*, or both.

Lithic Borohemists are like Typic Borohemists except for *f*, with or without all or any of *a*, *b*, *c*, or *e*.

Sapric Borohemists are like Typic Borohemists except for *b(2)*.

Sapric Terric Borohemists are like Typic Borohemists except for *b(2)* and *c*, with or without *a* or *e*, or both.

Terric Borohemists are like Typic Borohemists except for *c*, with or without all or any of *a*, *d*, or *e*.

Cryohemists

Distinctions between Typic Cryohemists and other subgroups

Typic Cryohemists are the Cryohemists that

a. Do not have a lithic contact within the control section;

b. Have a mean annual soil temperature higher than 0°C;

c. Do not have a mineral layer 30 cm or more thick that has its upper boundary in the control section below the surface tier;

d. Do not have a mineral layer between 5 and 30 cm thick within organic materials or do not have two or more thin, continuous mineral layers in the control section below the surface tier;

e. Have organic materials that are continuous laterally throughout each pedon in at least the surface tier.

Fluvaquentic Cryohemists are like Typic Cryohemists except for *d*.

Lithic Cryohemists are like Typic Cryohemists except for *a* or for *a* and *b*.

Pergelic Cryohemists are like Typic Cryohemists except for *b*, with or without *d*.

Terric Cryohemists are like Typic Cryohemists except for *c*, with or without *d*.

Luvihemists

Luvihemists are not known to occur in the United States but the great group is provided tentatively for use in other countries if needed. They are the Hemists that have a horizon that is 2 cm or more thick within the control section, and half or more of the volume of that horizon is humilluvic materials. Because these soils cannot be studied in the United States, a precise definition is not attempted here. It should be noted, however, that they are normally acid and that they have been cultivated for a long time.

Medihemists

Distinctions between Typic Medihemists and other sub-groups

Typic Medihemists are the Medihemists that
a. Do not have a mineral layer between 5 and 30 cm thick within organic materials or do not have two or more thin, continuous mineral layers in the control section below the surface tier;
b. Have
    (1) Less than 25 cm of the subsurface and bottom tiers consisting of fibric materials; and
    (2) Less than 25 cm of the subsurface and bottom tiers consisting of sapric materials;
c. Do not have a mineral layer 30 cm or more thick that has its upper boundary in the control section below the surface tier;
d. Do not have a layer of water within the control section beneath the surface tier;
e. Do not have limnic layer(s) that are 5 cm or more thick within the control section; and
f. Do not have a lithic contact within the control section.
Fibric Medihemists are like Typic Medihemists except for *b(1)*.
Fibric Terric Medihemists are like Typic Medihemists except for *b(1)* and *c*, with or without *a* or *e*, or both.
Fluvaquentic Medihemists are like Typic Medihemists except for *a*, with or without *b*.
Hydric Medihemists are like Typic Medihemists except for *d*.
Limnic Medihemists are like Typic Medihemists except for *e*, with or without *a* or *b*, or both.
Lithic Medihemists are like Typic Medihemists except for *f*, with or without all or any of *a*, *b*, *c*, or *e*.
Sapric Medihemists are like Typic Medihemists except for *b(2)*.
Sapric Terric Medihemists are like Typic Medihemists except for *b(2)* and *c*, with or without *a* or *e*, or both.
Terric Medihemists are like Typic Medihemists except for *c*, with or without *a* or *e*, or both.

Sulfihemists

Distinctions between Typic Sulfihemists and other subgroups

Typic Sulfihemists are the Sulfihemists that
a. Do not have a mineral layer 30 cm or more thick that has its lower boundary in the control section below the surface tier.
Terric Sulfihemists are like Typic Sulfihemists except for *a*.

Sulfohemists

Sulfohemists are the Hemists that have a sulfuric horizon whose upper boundary is within 50 cm of the surface.

The Sulfohemists are rare in the world and, provisionally, all Sulfohemists are considered to be Typic Sulfohemists.

Tropohemists

Distinctions between Typic Tropohemists and other subgroups

Typic Tropohemists are the Tropohemists that
a. Do not have a mineral layer between 5 and 30 cm thick within organic materials or do not have two or more thin, continuous mineral layers in the control section below the surface tier;
b. Have
  (1) Less than 25 cm of the subsurface and bottom tiers consisting of fibric materials; and
  (2) Less than 25 cm of the subsurface and bottom tiers consisting of sapric materials;
c. Do not have a mineral layer 30 cm or more thick that has its upper boundary in the control section below the surface tier;
d. Do not have a layer of water within the control section beneath the surface tier;
e. Do not have a limnic layer(s) that are 5 cm or more thick within the control section; and
f. Do not have a lithic contact within the control section.
Fibric Tropohemists are like Typic Tropohemists except for *b(1)*.
Fibric Terric Tropohemists are like Typic Tropohemists except for *b(1)* and *c*, with or without *a*.
Fluvaquentic Tropohemists are like Typic Tropohemists except for *a*, with or without *b*.
Hydric Tropohemists are like Typic Tropohemists except for *d*.
Limnic Tropohemists are like Typic Tropohemists except for *e*, with or without *a* or *b*, or both.
Lithic Tropohemists are like Typic Tropohemists except for *f*, with or without all or any of *a*, *b*, *c*, or *e*.
Sapric Tropohemists are like Typic Tropohemists except for *b(2)*.
Sapric Terric Tropohemists are like Typic Tropohemists except for *b(2)* and *c*, with or without *a* or *e*, or both.
Terric Tropohemists are like Typic Tropohemists except for *c*, with or without *a* or *e*, or both.

Saprists

Key to great groups

ADA. Saprists that are frozen in some layer within the control section about 2 months after the summer solstice or that are never frozen below a depth of 5 cm but have a mean annual soil temperature lower than 8°C.

Cryosaprists, p. 142

ADB. Other Saprists that have a mean annual soil temperature lower than 8°C.

Borosaprists, p. 141

ADC. Other Saprists that have <5°C difference between summer and mean winter soil temperatures at a depth of 30 cm.

Troposaprists, p. 143

ADD. Other Saprists that do not have a horizon of humilluvic materials 2 cm or more thick.

Medisaprists, p. 142

In addition to the great groups listed in the key, a great group of Vermisaprists may be needed, particularly for soils that have been drained for a long time, but a definition cannot be suggested at present.

Borosaprists

Distinctions between Typic Borosaprists and other subgroups

Typic Borosaprists are the Borosaprists that
a. Do not have a mineral layer between 5 and 30 cm thick with organic materials or do not have two or more thin, continuous mineral layers in the control section below the surface tier;
b. Have
    (1) Less than 12.5 cm of the subsurface and bottom tiers consisting of fibric materials; and
    (2) Less than 25 cm of the subsurface and bottom tiers consisting of hemic materials;
c. Do not have a mineral layer 30 cm or more thick that has its upper boundary in the control section below the surface tier;
d. Do not have a layer of water within the control section beneath the surface tier;
e. Do not have limnic layer(s) that are 5 cm or more thick within the control section; and
f. Do not have a lithic contact within the control section.
Fibric Borosaprists are like Typic Borosaprists except for *b(1)*.
Fibric Terric Borosaprists are like Typic Borosaprists except for *b(1)* and *c*, with or without *a* or *e*, or both.
Fluvaquentic Borosaprists are like Typic Borosaprists except for *a*, with or without *b*.
Hemic Borosaprists are like Typic Borosaprists except for *b(2)*.
Hemic Terric Borosaprists are like Typic Borosaprists except for *b(2)* and *c*, with or without *a* or *e*, or both.
Limnic Borosaprists are like Typic Borosaprists except for *e*, with or without *a* or *b*, or both.
Lithic Borosaprists are like Typic Borosaprists except for *f*, with or without all or any of *a*, *b*, *c*, or *e*.
Terric Borosaprists are like Typic Borosaprists except for *c*, with or without *a* or *e*, or both.

Cryosaprists

Distinctions between Typic Cryosaprists and other subgroups

Typic Cryosaprists are the Cryosaprists that
a. Do not have a lithic contact within the control section;
b. Have a mean annual soil temperature higher than 0°C;
c. Do not have a mineral layer 30 cm or more thick that has its upper boundary in the control section below the surface tier;
d. Do not have a mineral layer between 5 and 30 cm thick within organic materials or do not have two or more thin, continuous mineral layers in the control section below the surface tier; and
e. Have organic materials that are continuous laterally in at least the surface tier throughout each pedon.
Fluvaquentic Cryosaprists are like Typic Cryosaprists except for *d*.
Lithic Cryosaprists are like Typic Cryosaprists except for *a* or for *a* and *b*.
Pergelic Cryosaprists are like Typic Cryosaprists except for *b* or for *b* and *d*.
Terric Cryosaprists are like Typic Cryosaprists except for *c*.

Medisaprists

Distinctions between Typic Medisaprists and other subgroups

Typic Medisaprists are the Medisaprists that
a. Do not have a mineral layer between 5 and 30 cm thick within organic materials or do not have two or more thin, continuous mineral layers in the control section below the surface tier;
b. Have
    (1) Less than 12.5 cm of the subsurface and bottom tiers consisting of fibric materials; and
    (2) Less than 25 cm of the subsurface and bottom tiers consisting of hemic materials;
c. Do not have a mineral layer 30 cm or more thick that has its upper boundary in the control section below the surface tier;
d. Do not have a layer of water within the control section beneath the surface tier;
e. Do not have limnic layer(s) that are 5 cm or more thick within the control section; and
f. Do not have a lithic contact within the control section.
Fibric Medisaprists are like Typic Medisaprists except for *b(1)*.
Fibric Terric Medisaprists are like Typic Medisaprists except for *b(1)* and *c*, with or without *a* or *e*, or both.
Fluvaquentic Medisaprists are like Typic Medisaprists except for *a*, with or without *b*.
Hemic Medisaprists are like Typic Medisaprists except for *b(2)*.
Hemic Terric Medisaprists are like Typic Medisaprists except for *b(2)* and *c*, with or without *a* or *e*, or both.
Limnic Medisaprists are like Typic Medisaprists except for *e*, with or without *a* or *b*, or both.
Lithic Medisaprists are like Typic Medisaprists except for *f*, with or without all or any of *a*, *b*, *c*, or *e*.

Terric Medisaprists are like Typic Medisaprists except for *c*, with or without *a* or *e*, or both.

Troposaprists

Distinctions between Typic Troposaprists and other subgroups

Typic Troposaprists are the Troposaprists that
a. Do not have a mineral layer between 5 and 30 cm thick within organic materials or do not have two or more thin, continuous mineral layers in the control section below the surface tier;
b. Have
   (1) Less than 12.5 cm of the subsurface and bottom tiers consisting of fibric materials; and
   (2) Less than 25 cm of the subsurface and bottom tiers consisting of hemic materials;
c. Do not have a mineral layer 30 cm or more thick that has its upper boundary in the control section below the surface tier;
d. Do not have a layer of water within the control section beneath the surface tier;
e. Do not have limnic layer(s) that are 5 cm or more thick within the control section; and
f. Do not have a lithic contact within the control section.
Fibric Troposaprists are like Typic Troposaprists except for *b(1)*.
Fibric Terric Troposaprists are like Typic Troposaprists except for *b(1)* and *c*, with or without *a* or *e*, or both.
Fluvaquentic Troposaprists are like Typic Troposaprists except for *a*, with or without *b*.
Hemic Troposaprists are like Typic Troposaprists except for *b(2)*.
Hemic Terric Troposaprists are like Typic Troposaprists except for *b(2)* and *c*, with or without *a* or *e*, or both.
Limnic Troposaprists are like Typic Troposaprists except for *e*, with or without *a* or *b*, or both.
Lithic Troposaprists are like Typic Troposaprists except for *f*, with or without all or any of *a*, *b*, *c*, or *e*.
Terric Troposaprists are like Typic Troposaprists except for *c*, with or without *a* or *e*, or both.

Chapter 9

Inceptisols

Key To Suborders

IA. Inceptisols that

    1. Have an aquic moisture regime or are artificially drained and have one or more of the following:

        a. A histic epipedon;

        b. A sulfuric horizon that has its upper boundary within 50 cm of the mineral soil surface;

        c. An umbric or mollic epipedon that is underlain immediately or at a depth <50 cm below the soil surface by a horizon that has dominant colors, moist, on ped faces, or in the matrix if peds are absent, as follows:

            (1) If there is mottling, chroma is 2 or less[1];

            (2) If there is no mottling, chroma is 1 or less;

        d. An ochric epipedon that is underlain at a depth <50 cm below the mineral soil surface by a cambic horizon or a fragipan either or both of which has dominant color, moist, on ped faces, or in the matrix if peds are absent as follows:

            (1) If there is mottling, chroma is 2 or less[2];

            (2) If there is no mottling, chroma is 1 or less;

    2. Or have an SAR $\geq$13 (or sodium saturation that is 15 percent or more) in half or more of the soil to a depth of 50 cm that decreases with depth below 50 cm and ground water within 1 m of the surface at some time of the year.

Aquepts, p. 149

IB. Other Inceptisols that have to a depth of 35 cm or more, or to a lithic or paralithic contact if one is shallower than 35 cm, one or both of the following:

    1. Bulk density (at 33 kPa water retention) of the fine-earth fraction that is <0.85 g per cubic centimeter and an exchange complex that is dominated by amorphous materials; or

    2. Sixty percent or more of the soil (by weight) is vitric[3] volcanic ash, cinders or other pyroclastic materials.

Andepts, p. 146

IC. Other Inceptisols that have a plaggen epipedon.

Plaggepts, p. 163

ID. Other Inceptisols that have an isomesic or warmer iso temperature regime.

Tropepts, p. 163

---

[1] If the hue is redder than 10YR because of red parent materials that remain red after citrate-dithionite extraction, the requirement for low chroma is waived.

[2] If the hue is redder than 10YR because of red parent materials that remain red after citrate-dithionite extraction, the requirement for low chroma is waived.

[3] Included in the meaning of vitric materials in this definition are crystalline particles that are coated with glass and partially devitrified glass as well as glass.

IE. Other Inceptisols that have an ochric epipedon; or that have an umbric or mollic epipedon that is <25 cm thick and have also a mesic or warmer soil temperature regime.

Ochrepts, p. 156

IF. Other Inceptisols.

Umbrepts, p. 167

Andepts

Key to great groups

IBA. Andepts that have a cryic or pergelic temperature regime.

Cryandepts, p. 146

IBB. Other Andepts that have a duripan that has its upper boundary within 1 m of the soil surface.

Durandepts, p. 147

IBC. Other Andepts that have clays that dehydrate irreversibly into aggregates of sand and gravel size.

Hydrandepts, p. 149

IBD. Other Andepts that have a placic horizon within 1 m of the soil surface in half or more of each pedon.

Placandepts, p. 149

IBE. Other Andepts that are not thixotropic and in which the weighted average 1500 kPa water retention[4] of the fine-earth fraction
is <20 percent for all horizons between a depth of 25 cm and 1 m or between 25 cm and a lithic or paralithic contact if one is shallower than 1 m.

Vitrandepts, p. 149

IBF. Other Andepts that have a base saturation (by NH_4OAc) of 50 percent or more in some subhorizon between a depth of 25 and 75 cm.

Eutrandepts, p. 148

IBG. Other Andepts.

Dystrandepts, p. 147

Cryandepts

Distinctions between Typic Cryandepts and other subgroups

The subgroups of Cryandepts have not been fully developed because the soils are too few and those in the United States are too inaccessible. To date they have had only preliminary study.

---

[4] The amount of water retained at a tension of 1500 kPa may be reduced by drying these soils. Since 1500 kPa water retention is used as a measure of effective clay content, it would be unrealistic to use the extremely high value obtained on a field-moist Andept. That value reflects mostly the climatic history rather than a basic soil property. The value for 1500 kPa water retention that is referred to here, therefore, is that of a sample that has been dried at 40°C.

Typic Cryandepts are Cryandepts that
a. Do not have mottles that have chroma of 2 or less within 1 m of the soil surface;
b. Are not thixotropic in half or more of the thickness of all horizons between depths of 25 cm and 1 m, and the weighted average 1500 kPa water retention[5] is <20 percent between depths of 25 cm and 1 m, or between 25 cm and a lithic or paralithic contact if there is one between 50 cm and 1 m;
c. Do not have a lithic contact within 50 cm of the soil surface;
d. Have an epipedon that has the color and thickness of a mollic epipedon or have a horizon that meets the color and thickness requirements of a mollic epipedon and has its upper boundary within 50 cm of the soil surface;
e. Have a mean annual soil temperature higher than 0°C;
f. Do not have a placic horizon;
g. Do not have a duripan that has its upper boundary within 1 m of the soil surface.
Dystric Cryandepts are like Typic Cryandepts except for *b*, with or without *d*, and base saturation (by NH_4OAc) is <50 percent in all subhorizons between 25 and 75 cm.
Dystric Lithic Cryandepts are like Typic Cryandepts except for *b* and *c*, and base saturation (by NH_4OAc) is <50 percent in all subhorizons between 25 and 75 cm.
Entic Cryandepts are like Typic Cryandepts except for *d*.
Lithic Cryandepts are like Typic Cryandepts except for *c* or for *c* and *d*.

Durandepts

Proposed distinctions between Typic Durandepts and other subgroups

Typic Durandepts are the Durandepts that
a. Have an ustic moisture regime;
b. Have an epipedon that meets the color and thickness requirements of a mollic epipedon;
c. Have a duripan that is continuous throughout each pedon or, if fractured, the average lateral dimensions of the fragments are ≥10 cm.
Entic Durandepts are like Typic Durandepts except for *c*.
Xeric Durandepts are like Typic Durandepts except for *a*, and they have a xeric moisture regime.

Dystrandepts

Distinctions between Typic Dystrandepts and other subgroups

Typic Dystrandepts are the Dystrandepts that
a. Do not have mottles that have chroma of 2 or less within 1 m of the soil surface;

---

[5]The amount of water retained at a tension of 1500 kPa may be reduced by drying these soils. Since 1500 kPa water retention is used as a measure of effective clay content, it would be unrealistic to use the extremely high value obtained on a field-moist Andept. That value reflects mostly the climatic history rather than a basic soil property. The value for 1500 kPa water retention that is referred to here, therefore, is that of a sample that has been dried at 40°C.

b. Have an epipedon that is 25 cm or more thick and meets
the color requirements of a mollic epipedon;
c. Are not thixotropic in any horizon between depths of 25
cm and 1 m;
d. Do not have a lithic contact within 50 cm of the soil
surface;
e. Have cation-exchange capacity[6] of >30 cmol(+) kg^{-1} soil
(by NH$_4$OAc) in all subhorizons above a lithic contact or a
depth of 1 m, whichever is shallower, or have >10 percent
weatherable minerals in the 20- to 200-micrometer fraction.
Aquic Dystrandepts are like Typic Dystrandepts except for *a.*
Entic Dystrandepts are like Typic Dystrandepts except for *b.*
Hydric Dystrandepts are like Typic Dystrandepts except for
c, with or without *b.*
Hydric Lithic Dystrandepts are like Typic Dystrandepts
except for *c* and *d*, with or without *b.*
Lithic Dystrandepts are like Typic Dystrandepts except for
d, with or without *b.*
Oxic Dystrandepts are like Typic Dystrandepts except for *e*,
with or without *b*, and they have <10 percent weatherable
minerals in the 20- to 200-micrometer fraction.

Eutrandepts

Distinctions between Typic Eutrandepts and other subgroups

Typic Eutrandepts are the Eutrandepts that
a. Do not have mottles that have chroma of 2 or less within
1 m of the soil surface;
b. Have an epipedon, 25 cm or more thick, that meets the
color requirements of a mollic epipedon;
c. Do not have a lithic contact within 50 cm of the soil
surface;
d. Do not have a subhorizon within 1.5 m of the surface
that contains soft, powdery secondary lime;
e. Have an ustic moisture regime;
f. Do not have a horizon within 1 m of the surface that is
>15 cm thick and that either contains 20 percent or more
(by volume) durinodes or has >50 percent (by volume)
fragments of a duripan in which the average horizontal
repeat distance between vertical cracks is <10 cm.
Duric Eutrandepts are like Typic Eutrandepts except for *f.*
Entic Eutrandepts are like Typic Eutrandepts except for *b* or
for *b* and *d.*
Lithic Eutrandepts are like Typic Eutrandepts except for *c.*
Udic Eutrandepts are like Typic Eutrandepts except for *e*,
and they have a udic moisture regime.
Ustollic Eutrandepts are like Typic Eutrandepts except for *d.*
Xeric Eutrandepts are like Typic Eutrandepts except for *e*,
and they have a xeric moisture regime.

---

[6]Andepts in a perhumid climate are never dry while they remain in place.
They lose CEC if allowed to become air dry before the determination is made.
The values used in these definitions are those determined on soil samples that
have never been allowed to dry, but the values have been recalculated to an
ovendry basis. The CEC of samples that have been air dried may be used if it
has been determined that drying does not affect the CEC of that soil.

Hydrandepts

Distinction between Typic and Lithic Hydrandepts

Typic Hydrandepts are the Hydrandepts that
a. Do not have a lithic contact within 50 cm of the soil surface.
Lithic Hydrandepts are like Typic Hydrandepts except for *a*.

Placandepts

These are the Andepts that have a placic horizon. Like the Hydrandepts, they are mostly in very humid climates that do not have a dry season. The placic horizon is a barrier to water movement and root development. Water commonly saturates much of the soil above the pan for variable periods. This great group is represented by a very few soil series in the United States. The soils have been reported in other countries. These soils were called Hydrol Humic Latosols in the 1938 classification as modified in 1955.

Subgroups have not been developed because the soils are too poorly represented in the United States. It is thought that soils of the typic subgroup should have, within 1 m of the surface, a placic horizon that is continuous through each pedon.

Vitrandepts

Distinctions between Typic Vitrandepts and other subgroups

Typic Vitrandepts are the Vitrandepts that
a. Do not have mottles that have chroma of 2 or less within 1 m of the soil surface if the mottled horizon is saturated with water in most years at some period when its temperature is >5°C or if the soil is artificially drained;
b. Do not have a lithic contact within 50 cm of the soil surface; and
c. Have an ochric epipedon.
Aquic Vitrandepts are like Typic Vitrandepts except for *a*.
Lithic Vitrandepts are like Typic Vitrandepts except for *b*.
Lithic Mollic Vitrandepts are like Typic Vitrandepts except for *b* and *c*, and they have a mollic epipedon.
Lithic Umbric Vitrandepts are like Typic Vitrandepts except for *b* and *c*, and they have an umbric epipedon.
Mollic Vitrandepts are like Typic Vitrandepts except for *c*, and they have a mollic epipedon.
Plaggic Vitrandepts are like Typic Vitrandepts except for *c*, and they have either a plaggen epipedon or an epipedon that meets all requirements for a plaggen epipedon except thickness and is 30 cm or more thick.
Umbric Vitrandepts are like Typic Vitrandepts except for *c*, and they have an umbric epipedon.

Aquepts

Key to great groups

IAA. Aquepts that have a sulfuric horizon whose upper boundary is within 50 cm of the mineral soil surface.
Sulfaquepts, p. 155

IAB. Other Aquepts that have a placic horizon within 1 m
of the mineral soil surface in half or more of each pedon.
Placaquepts, p. 154

IAC. Other Aquepts that have an SAR $\geq$13 (or have sodium
saturation that is $\geq$15 percent) in half or more of the upper
50 cm of soil that decreases with depth below 50 cm.
Halaquepts, p. 152

IAD. Other Aquepts that have a fragipan.
Fragiaquepts, p. 152

IAE. Other Aquepts that have a cryic or pergelic soil
temperature regime.
Cryaquepts, p. 151

IAF. Other Aquepts that have plinthite that forms a
continuous phase or constitutes more than half the matrix
within some subhorizon in the upper 1.25 m of the soil.
Plinthaquepts, p. 154

IAG. Other Aquepts that have, to a depth of 35 cm or more
or to a lithic or paralithic contact if one is shallower than 35
cm, one or both of the following:
1. Bulk density (at 33 kPa water retention) of the fine-earth
fraction that is <0.85 g per cubic centimeter and an
exchange complex that is dominated by amorphous materials;
or
2. Sixty percent or more of the soil (by weight) is vitric[7]
volcanic ash, cinders, or other pyroclastic materials.
Andaquepts, p. 150

IAH. Other Aquepts that have a difference of <5°C between
the mean summer and mean winter soil temperatures at a
depth of 50 cm or at a lithic or paralithic contact, whichever
is shallower.
Tropaquepts, p. 155

IAI. Other Aquepts that have an umbric, a mollic, or a
histic epipedon.
Humaquepts, p. 154

IAJ. Other Aquepts.
Haplaquepts, p. 152

Andaquepts

Distinctions between Typic Andaquepts and other subgroups

Typic Andaquepts are the Andaquepts that
a. Have in 60 percent or more of the matrix in all
subhorizons between the A or Ap horizon and a depth of 75
cm one or more of the following:
(1) If mottled and if the hue is 2.5Y or redder and the
value, moist, is 5 or more, the chroma, moist, is 2 or less;
(2) If mottled and if the hue is yellower than 2.5Y, the
chroma, moist, is 2 or less;

---

[7]Included in the meaning of vitric materials in this definition are crystalline
particles that are coated with glass and partially devitrified glass as well as
glass.

(3) Whether mottled or not, the chroma, moist, is 1 or less; and
b. Have an umbric epipedon.
Aeric Andaquepts are like Typic Andaquepts except for *a*.
Aeric Mollic Andaquepts are like Typic Andaquepts except for *a* and *b*, and they have a mollic epipedon.
Haplic Andaquepts are like Typic Andaquepts except for *b*, and they have an ochric epipedon.
Mollic Andaquepts are like Typic Andaquepts except for *b*, and they have a mollic epipedon.

Cryaquepts

Distinctions between Typic Cryaquepts and other defined subgroups

It seems probable that a number of subgroups besides those here defined will be needed when the soils that occur in striped or polygonal patterns have been studied in more detail.

Typic Cryaquepts are the Cryaquepts that
a. Have chroma of 2 or less in 60 percent or more of the mass of all horizons between depths of 15 and 50 cm;
b. Do not have a layer in the upper 75 cm that has texture finer than loamy fine sand, that is as much as 18 cm thick, that has a bulk density (at 33 kPa tension) of 0.95 g per cubic centimeter or less in the fine-earth fraction and that has either (1) a ratio of measured clay to 1500 kPa water (percentages) of 1.25 or less, or (2) a ratio of CEC (at pH near 8) to 1500 kPa water of >1.5 and more exchange acidity than the sum of bases plus KCl-extractable aluminum;
c. Do not have a histic epipedon;
d. Do not have an umbric or a mollic epipedon;
e. Do not have a lithic contact within 50 cm of the soil surface;
f. Have a mean annual soil temperature that is higher than 0°C.
Aeric Cryaquepts are like Typic Cryaquepts except for *a*.
Aeric Humic Cryaquepts are like Typic Cryaquepts except for *a* and *d*, and they have an umbric epipedon.
Andic Cryaquepts are like Typic Cryaquepts except for *b* or for *b* and *d*.
Histic Cryaquepts are like Typic Cryaquepts except for *c* or for *c* and *d*.
Histic Lithic Cryaquepts are like Typic Cryaquepts except for *c* and *e*, with or without *f*.
Histic Pergelic Cryaquepts are like Typic Cryaquepts except for *c* and *f*, and the histic epipedon is continuous in each pedon.
Humic Cryaquepts are like Typic Cryaquepts except for *d*, and they have an umbric epipedon.
Humic Pergelic Cryaquepts are like Typic Cryaquepts except for *d* and *f*, with or without *a*, and they have an umbric epipedon.
Pergelic Cryaquepts are like Typic Cryaquepts except for *f* or for *f* and *a*.
Pergelic Ruptic-Histic Cryaquepts are like Typic Cryaquepts except for *c* and *f*, and the histic epipedon is not continuous in each pedon.

Fragiaquepts

Distinctions between Typic Fragiaquepts and other
subgroups

Typic Fragiaquepts are the Fragiaquepts that
a. Do not have a histic, mollic, or umbric epipedon; and
b. Have, in 60 percent or more of the matrix of all
subhorizons between the plow layer or, if there is no plow
layer, a depth of 15 cm and a depth of 75 cm or more,
moist colors as follows:
    (1) If there is mottling, chroma of 2 or less;
    (2) If there is no mottling, chroma of 1 or less.
Aeric Fragiaquepts are like Typic Fragiaquepts except for *b.*
Humic Fragiaquepts are like Typic Fragiaquepts except for
a.

Halaquepts

Distinctions between Typic Halaquepts and other subgroups

Typic Halaquepts are the Halaquepts that
a. Have chroma of 2 or less and a hue of 5Y or redder in 60
percent or more of the matrix in all subhorizons between
depths of 15 and 75 cm;
b. Have an ochric epipedon;
c. Do not have the following combination of characteristics:
    (1) Cracks at some period in most years that are 1 cm or
    more wide at a depth of 50 cm, that are at least 30 cm
    long in some part, and that extend upward to the soil
    surface or to the base of an Ap horizon;
    (2) A coefficient of linear extensibility (COLE) of 0.09
    or more in a horizon or horizons at least 50 cm thick and
    a potential linear extensibility of 6 cm or more in the
    upper 1 m of the soil or in the whole soil if a lithic or
    paralithic contact is deeper than 50 cm but shallower than
    1 m; and
    (3) More than 35 percent clay in horizons that total >50
    cm in thickness.
Aeric Halaquepts are like Typic Halaquepts except for *a.*
Mollic Halaquepts are like Typic Halaquepts except for *b.*
Vertic Halaquepts are like Typic Halaquepts except for *c.*

Haplaquepts

Distinctions between Typic Haplaquepts and other subgroups

Typic Haplaquepts are the Haplaquepts that
a. Have, in 60 percent or more of the matrix in all
subhorizons between the Ap horizon and a depth of 75 cm,
one or more of the following:
    (1) If mottled and the mean annual soil temperature is
    lower than 15°C, moist chroma of 2 or less;
    (2) If mottled and the mean annual soil temperature is
    15°C or higher:
        (a) If the hue is 2.5Y or redder[8] and the value, moist,
        is more than 5, the chroma, moist, is 2 or less;

---

[8] If the hue is 7.5YR or redder in the matrix and if peds are present, the ped
exteriors should have dominant chroma, moist, of 1 or less and the ped interiors should have mottles that have chroma, moist, of 2 or less; if there are no

(b) If the hue is 2.5Y or redder and the value, moist, is 5 or less, the chroma, moist, is 1 or less;
(c) If the hue is yellower than 2.5Y, the chroma, moist, is 2 or less;
(3) The chroma, moist, is 1 or less and mottles may or may not be present;

b. Do not have a layer in the upper 75 cm that has texture finer than loamy fine sand, that is as much as 18 cm thick, that has bulk density (at 33 kPa water tension) of 0.95 g per cubic centimeter or less in the fine-earth fraction, and that has either of the following:

(1) A ratio of measured clay to 1500 kPa water (percentages) of 1.25 or less; or
(2) A ratio of CEC (at pH near 8) to 1500 kPa water of >1.5 and more exchange acidity than the sum of bases plus KCl-extractable aluminum;

c. Have an Ap horizon that has a color value, moist, of 4 or more or has a value, dry, of 6 or more when crushed and smoothed, or have an A horizon that is <15 cm thick if its color value, moist, is lower than 3.5;

d. Have an *n* value of <0.9 between depths of 50 and 80 cm and <0.7 in all layers between 20 and 50 cm;

e. Do not have a lithic contact within 50 cm of the soil surface;

f. Do not have the following combination of characteristics:

(1) Cracks at some period in most years that are 1 cm or more wide at a depth of 50 cm, that are at least 30 cm long in some part, and that extend upward to the soil surface or to the base of an Ap horizon; and
(2) A coefficient of linear extensibility (COLE) of 0.09 or more in a horizon or horizons at least 50 cm thick and a potential linear extensibility of 6 cm or more in the upper 1 m of the soil or in the whole soil if a lithic or paralithic contact is deeper than 50 cm but shallower than 1 m; and
(3) More than 35 percent clay in horizons that total >50 cm in thickness; and

g. Do not have either of the following:

(1) Jarosite mottles and a pH between 3.5 and 4.0 (1:1 water, air dried slowly in shade) in some subhorizon within 50 cm of the soil surface, or
(2) Jarosite mottles and a pH <4.0 (1:1 water, air dried slowly in shade) in some subhorizon between depths of 50 and 150 cm.

Aeric Haplaquepts are like Typic Haplaquepts except for *a* or for *a* and *c*.

Humic Haplaquepts are like Typic Haplaquepts except for *c*, and the base saturation (by NH_4OAc) is <50 percent in some horizon and does not increase with depth to a value of 50 percent or more.

Lithic Haplaquepts are like Typic Haplaquepts except for *e* or for *a* and *e*.

Mollic Haplaquepts are like Typic Haplaquepts except for *c*, and the base saturation (by NH_4OAc) is 50 percent or more throughout or increases with depth to a value of 50 percent or more.

Sulfic Haplaquepts are like Typic Haplaquepts except for *g*, with or without all or any of *a*, *c*, *d*, or *f*.

INC

---

peds, the chroma, moist, should be 1 or less immediately below any surface horizon that has a value, moist, less than 5.

Vertic Haplaquepts are like Typic Haplaquepts except for *f*, with or without *a* or *c*, or both.

Humaquepts

Distinctions between Typic Humaquepts and other subgroups

Typic Humaquepts are the Humaquepts that
a. Have chroma of 2 or less, moist, and hue of 5Y or redder in 60 percent or more of the matrix in all subhorizons between depths of 15 and 75 cm;
b. Have mottles or have iron-manganese concretions within a depth 30 cm below the base of the epipedon if the chroma within that depth is 1 or more, the hue is redder than 5Y, and the value, moist, is 5 or more;
c. Do not have a layer in the upper 75 cm that has texture finer than loamy fine sand, that is as much as 18 cm thick, that has a bulk density (at 33 kPa tension) of 0.95 g per cubic centimeter or less in the fine-earth fraction, and that has either
    (1) a ratio of measured clay to 1500 kPa water (percentages) of 1.25 or less, or
    (2) a ratio of CEC (at pH near 8) to 1500 kPa water of >1.5 and more exchange acidity than the sum of bases plus KCl-extractable aluminum;
d. Have an epipedon that is <60 cm thick;
e. Have a content of organic carbon that decreases regularly with depth and, unless a lithic or a paralithic contact occurs at a shallower depth, reaches a level of 0.2 percent or less within 1.25 m of the soil surface;
f. Do not have a histic epipedon whose upper boundary is at or near the soil surface; and
g. Have an *n* value of <0.9 between depths of 50 and 80 cm and of 0.7 or less in all layers between depths of 20 and 50 cm.
Cumulic Humaquepts are like Typic Humaquepts except for *d* and *e*.
Fluvaquentic Humaquepts are like Typic Humaquepts except for *e*.
Histic Humaquepts are like Typic Humaquepts except for *f*, with or without all or any of *a*, *b*, or *e*.

Placaquepts

Distinctions between Typic Placaquepts and other subgroups

Typic Placaquepts are the Placaquepts that
a. Do not have a histic epipedon; and
b. Have a continuous placic horizon within 1 m of the soil surface throughout each pedon.
Histic Placaquepts are like Typic Placaquepts except for *a*.

Plinthaquepts

These are mainly Aquepts of intertropical regions. They have plinthite that forms a continuous phase or occupies more than half the matrix of some subhorizon deeper than 30 cm but within 1.25 m of the soil surface. These are soils in which the ground-water level fluctuates appreciably during the year. Water is at or near the surface during the rainy season but drops during a dry season. Most of these soils are in relatively recent alluvium, probably of late-Pleistocene or Holocene age. Weatherable minerals are present in appreciable amounts. These soils are not known to occur in the United States, but the great group is provided because the soils are thought to be extensive in parts of the Amazon basin.

Sulfaquepts

Sulfaquepts are the Aquepts that have a sulfuric horizon that has its upper boundary within 50 cm of the soil surface.

Subgroups of Sulfaquepts have not been fully developed. It is thought that few subgroups are needed because the sulfuric horizon is so highly toxic. The Typic Sulfaquepts have a sulfuric horizon within 50 cm of the soil surface. A histic epipedon is permitted but is not required.

Tropaquepts

Distinctions between Typic Tropaquepts and other subgroups

Typic Tropaquepts are the Tropaquepts that
a. Have in 60 percent or more of the matrix in all subhorizons between the A or Ap horizon and a depth of 75 cm one or more of the following:
  (1) If mottled and if the hue is 2.5Y or redder and the value, moist, is >5, the chroma, moist, is 2 or less; if the value, moist, is 5 or less, the chroma, moist, is 1 or less;
  (2) If mottled and if the hue is yellower than 2.5Y, the chroma, moist, is 2 or less;
  (3) The chroma, moist, is 1 or less whether mottled or not;
b. Do not have a histic epipedon that has its upper boundary at or near the surface;
c. Do not have a lithic contact within 50 cm of the soil surface;
d. Do not have the following combination of characteristics:
  (1) Cracks at some time in most years that are 1 cm or more wide at a depth of 50 cm, that are at least 30 cm long in some part, and that extend upward to the soil surface or to the base of an Ap horizon;
  (2) A coefficient of linear extensibility (COLE) of 0.09 or more in a horizon or horizons at least 50 cm thick and a potential linear extensibility of 6 cm or more in the upper 1 m of the soil or in the whole soil if a lithic or paralithic contact is deeper than 50 cm but shallower than 1 m; and
  (3) More than 35 percent clay in horizons that total >50 cm in thickness;
e. Have <5 percent (by volume) of plinthite in all subhorizons within 1.5 m of the soil surface;
f. Do not have either of the following:
      (1) Jarosite mottles and a pH between 3.5 and 4.0 (1:1 water, air dried slowly in shade) in some subhorizon within 50 cm of the soil surface; or
      (2) Jarosite mottles and a pH <4.0 (1:1 water, air dried slowly in shade) in some subhorizon between depths of 50 and 150 cm.
Aeric Tropaquepts are like Typic Tropaquepts except for *a.*
Histic Tropaquepts are like Typic Tropaquepts except for *b.*

Lithic Tropaquepts are like Typic Tropaquepts except for *c*.
Plinthic Tropaquepts are like Typic Tropaquepts except for *e*
or for *a* and *e*.
Sulfic Tropaquepts are like Typic Tropaquepts except for *f*,
with or without all or any of *a*, *b*, or *d*.
Vertic Tropaquepts are like Typic Tropaquepts except for *d*,
with or without *a*.

Ochrepts

Key to great groups

IEA. Ochrepts that have a fragipan.

Fragiochrepts, p. 160

IEB. Other Ochrepts that have a duripan whose upper
boundary is within 1 m of the soil surface.

Durochrepts, p. 157

IEC. Other Ochrepts that have a cryic or pergelic
temperature regime.

Cryochrepts, p. 156

IED. Other Ochrepts that have an ustic moisture regime.

Ustochrepts, p. 160

IEE. Other Ochrepts that have a xeric moisture regime.

Xerochrepts, p. 161

IEF. Other Ochrepts that have one or both of the following:
1. Carbonates in the cambic horizon or in the C horizon
but within the soil; or
2. Base saturation (by NH_4OAc) that is 60 percent or
more in some subhorizon between depths of 25 and 75
cm below the soil surface.

Eutrochrepts, p. 158

IEG. Other Ochrepts.

Dystrochrepts, p. 157

Cryochrepts

Distinctions between Typic Cryochrepts and other subgroups

Typic Cryochrepts are the Cryochrepts that
a. Have a mean annual soil temperature higher than 0°C;
b. Do not have a lithic contact within 50 cm of the soil
surface;
c. Do not have mottles that have chroma of 2 or less within
75 cm of the soil surface if the mottled horizon is saturated
with water at some period when its temperature is $\geq$5°C or
the soil has artificial drainage.
d. Do not have a layer in the upper 75 cm that has texture
finer than loamy fine sand, that is as much as 18 cm thick,
that has bulk density (at 33 kPa water tension) of 0.95 g per
cubic centimeter or less in the fine-earth fraction, and that
has either (1) a ratio of measured clay to 1500 kPa water
(percentages) of 1.25 or less, or (2) a ratio of CEC (at pH
near 8) to 1500 kPa water of >1.5 and more exchange
acidity than the sum of bases plus KCl-extractable
aluminum;

e. Have base saturation (by NH_4OAc) that is 60 percent or more in some subhorizon within 75 cm of the surface; and
f. Do not have lamellae within 75 cm of the soil surface that meet all requirements for an argillic horizon except thickness.

Alfic Cryochrepts are like Typic Cryochrepts except for *f*.
Andic Cryochrepts are like Typic Cryochrepts except for *d* or for *d* and *e*.
Aquic Cryochrepts are like Typic Cryochrepts except for *c* or *c* and *e*.
Dystric Cryochrepts are like Typic Cryochrepts except for *e*.
Lithic Cryochrepts are like Typic Cryochrepts except for *b*, with or without *a* or *d*, or both.
Pergelic Cryochrepts are like Typic Cryochrepts except for *a* or for *a* and *c*.

Durochrepts

Distinctions between Typic Durochrepts and other subgroups

Typic Durochrepts are the Durochrepts that
a. Have a platy or massive indurated duripan;
b. Do not have distinct or prominent mottles within the upper 30 cm;
c. Have a xeric moisture regime; and
d. Have base saturation (by NH_4OAc) of 60 percent or more in some part of the soil between depths of 25 and 75 cm below the soil surface.

Dystric Entic Durochrepts are like Typic Durochrepts except for *a* and *d*.
Entic Durochrepts are like Typic Durochrepts except for *a*.

Dystrochrepts

Distinctions between Typic Dystrochrepts and other subgroups

Typic Dystrochrepts are the Dystrochrepts that
a. Do not have a layer in the upper 75 cm that has texture finer than loamy fine sand, that is as much as 18 cm thick, that has a bulk density (at 33 kPa water tension) of 0.95 g per cubic centimeter or less in the fine-earth fraction, and that has either
    (1) a ratio of measured clay to 1500 kPa water (percentages) of 1.25 or less or
    (2) a ratio of CEC (at pH near 8) to 1500 kPa water of >1.5 and more exchange acidity than the sum of bases plus KCl-extractable aluminum;
b. Do not have mottles that have chroma of 2 or less within 60 cm of the soil surface if the mottled horizon is saturated with water at a time when its temperature is 5°C or higher, or the soil has artificial drainage.
c. Have a content of organic carbon[9] that decreases regularly with depth and, unless a lithic or a paralithic contact occurs at a shallower depth, reaches a level of 0.2 percent or less within 1.25 m of the surface; or have slopes >25 percent;
d. Do not have a lithic contact within 50 cm of the soil surface;

---

[9]The carbon should be of Holocene age. It is not the intent to include fossil carbon from bedrock.

e. Do not have an argillic horizon in any part of the pedon; and

f. Have an Ap horizon that has a color value, moist, of 4 or more or a color value, dry, of 6 or more, crushed and smoothed, or the upper soil to a depth of 18 cm, after mixing, has these colors.

Andic Dystrochrepts are like Typic Dystrochrepts except for *a*.

Aquic Dystrochrepts are like Typic Dystrochrepts except for *b*.

Fluvaquentic Dystrochrepts are like Typic Dystrochrepts except for *b* and *c*.

Fluventic Dystrochrepts are like Typic Dystrochrepts except for *c*.

Fluventic Umbric Dystrochrepts are like Typic Dystrochrepts except for *c* and *f*.

Lithic Dystrochrepts are like Typic Dystrochrepts except for *d*.

Lithic Ruptic-Alfic Dystrochrepts are like Typic Dystrochrepts except for *d* and *e*, they have an argillic horizon in less than half of each pedon, and their base saturation (by sum of cations) in the subhorizon just above the lithic contact that is 35 percent or more.

Lithic Ruptic-Ultic Dystrochrepts are like Typic Dystrochrepts except for *d* and *e*, and they have an argillic horizon in less than half of each pedon and have base saturation (by sum of cations) in the subhorizon just above the lithic contact that is <35 percent.

Ruptic-Alfic Dystrochrepts are like Typic Dystrochrepts except for *e* and they have an argillic horizon in less than half of each pedon and base saturation (by sum of cations) that is 35 percent or more at a depth 1.25 m below the upper boundary of the argillic horizon or just above a lithic or paralithic contact if one is present at a shallower depth.

Ruptic-Ultic Dystrochrepts are like Typic Dystrochrepts except for *e*, and they have an argillic horizon in less than half of each pedon and have base saturation (by sum of cations) that is <35 percent at a depth 1.25 m below the top of the argillic horizon.

Umbric Dystrochrepts are like Typic Dystrochrepts except for *f*.

Eutrochrepts

Distinctions between Typic Eutrochrepts and other subgroups

Typic Eutrochrepts are the Eutrochrepts that
a. Do not have a layer in the upper 75 cm that has texture finer than loamy fine sand, that is as much as 18 cm thick, that has bulk density (at 33 kPa water tension) of 0.95 g per cubic centimeter or less in the fine-earth fraction and that has either of the following:
  (1) A ratio of measured clay to 1500 kPa water (percentages) of 1.25 or less; or
  (2) A ratio of CEC (at pH near 8) to 1500 kPa water of >1.5 and more exchange acidity than the sum of bases plus KCl-extractable aluminum;
b. Do not have mottles that have chroma of 2 or less within 60 cm of the soil surface if the mottled horizon is saturated

with water at some period when its temperature is $\geq5°C$ or
if the soil has artificial drainage;
c. Have texture of very fine sand or finer within 50 cm of
the soil surface;
d. Have carbonates within a depth of 1 m in some part of
each pedon;
e. Have a content of organic carbon that decreases regularly
with depth and, unless a lithic or a paralithic contact occurs
at a shallower depth, reaches a level of 0.2 percent or less
within 1.25 m of the soil surface; or have slopes >25
percent;
f. Do not have a lithic contact within 50 cm of the soil
surface in any part of the pedon;
g. Do not have an argillic horizon in any part of the pedon;
h. Have an ochric epipedon;
i. Do not have the following combination of characteristics:
    (1) Cracks at some period in most years that are 1 cm or
    more wide at a depth of 50 cm, that are at least 30 cm
    long in some part, and that extend to the soil surface or
    to the base of an Ap horizon,
    (2) A coefficient of linear extensibility (COLE) of 0.09
    or more in a horizon or horizons at least 50 cm thick and
    a potential linear extensibility of 6 cm or more in the
    upper 1 m of the soil or in the whole soil if a lithic or
    paralithic contact is deeper than 50 cm but shallower than
    1 m, and
    (3) More than 35 percent clay in horizons that total >50
    cm in thickness; or
j. Have <40 percent carbonates, including the coarse
fragments up to 75 mm in diameter, in and below the
cambic horizon but above a lithic or paralithic contact and
above a depth of 1 m.

Andic Dystric Eutrochrepts are like Typic Eutrochrepts
except for *a* and *d*.

Aquic Eutrochrepts are like Typic Eutrochrepts except for *b*.

Aquic Dystric Eutrochrepts are like Typic Eutrochrepts
except for *b* and *d*.

Arenic Eutrochrepts are like Typic Eutrochrepts except for *c*.

Dystric Eutrochrepts are like Typic Eutrochrepts except for
d.

Dystric Fluventic Eutrochrepts are like Typic Eutrochrepts
except for *d* and *e*.

Fluvaquentic Eutrochrepts are like Typic Eutrochrepts except
for *b* and *e*, with or without *d*.

Fluventic Eutrochrepts are like Typic Eutrochrepts except for
e.

Lithic Eutrochrepts are like Typic Eutrochrepts except for *f*
or for *d* and *f*.

Lithic Ruptic-Alfic Eutrochrepts are like Typic Eutrochrepts
except for *f* and *g*, and they have an argillic horizon in
some part but in less than half of each pedon.

Rendollic Eutrochrepts are like Typic Eutrochrepts except
for *j*.

Ruptic-Alfic Eutrochrepts are like Typic Eutrochrepts except
for *g* and have an argillic horizon in some part but in less
than half of each pedon.

Vertic Eutrochrepts are like Typic Eutrochrepts except for *i*,
with or without *e* or *b*, or both.

Fragiochrepts

Distinctions between Typic Fragiochrepts and other subgroups

Typic Fragiochrepts are the Fragiochrepts that
a. Do not have a layer in the upper 75 cm that has texture finer than loamy fine sand, that is as much as 18 cm thick, that has bulk density (at 33 kPa water tension) of 0.95 g per cubic centimeter or less in the fine-earth fraction, and that has either (1) a ratio of measured clay to 1500 kPa water (percentages) of 1.25 or less, or (2) a ratio of CEC (at pH near 8) to 1500 kPa water of >1.5 and more exchange acidity than the sum of bases plus KCl-extractable aluminum;
b. Do not have distinct or prominent mottles in the upper 30 cm of the soil; and
c. Have an ochric epipedon.
Andic Fragiochrepts are like Typic Fragiochrepts except for *a.*
Aquic Fragiochrepts are like Typic Fragiochrepts except for *b.*

Ustochrepts

Distinctions between Typic Ustochrepts and other subgroups

Typic Ustochrepts are the Ustochrepts that
a. Have a content of organic carbon[10] that decreases regularly with depth and, unless a lithic or a paralithic contact occurs at a shallower depth, reaches a level of 0.2 percent or less within 1.25 m of the soil surface; or have slopes >25 percent;
b. Do not have a lithic contact within 50 cm of the surface;
c. Do not have the following combination of characteristics:
  (1) Cracks at some period in most years that are 1 cm or more wide at a depth of 50 cm and that are at least 30 cm long in some part and that extend upward to the soil surface or the base of an Ap horizon,
  (2) A coefficient of linear extensibility (COLE) of 0.07 or more in a horizon or horizons at least 50 cm thick and a potential linear extensibility of 6 cm or more in the upper 1.25 m of the soil or in the whole soil if a lithic or paralithic contact is deeper than 50 cm but shallower than 1.25 m, and
  (3) More than 35 percent clay in horizons that total >50 cm in thickness;
d. Do not have mottles that have chroma of 2 or less within 75 cm of the soil surface if the mottled horizon is saturated with water at some period when its temperature is 5°C or more or if the soil has artificial drainage;
e. When neither irrigated nor fallowed to store moisture:
  (1) If the soil temperature regime is mesic or thermic, are dry for more than four-tenths of the cumulative days in some part of the moisture control section when the soil temperature at a depth of 50 cm exceeds 5°C; or
  (2) If the soil temperature regime is hyperthermic, isomesic, or warmer, the soils are dry in some or all parts

---

[10]The carbon should be of Holocene age. It is not the intent to include fossil carbon from bedrock.

of the moisture control section for more than 90 days
during a period when the soil temperature at a depth of
50 cm exceeds 8°C;
f. Do not have a layer in the upper 75 cm that has texture
finer than loamy fine sand, that is as much as 18 cm thick,
that has bulk density (at 33 kPa water tension) of 0.95 g per
cubic centimeter or less in the fine-earth fraction, and that
has either of the following:
(1) A ratio of measured clay to 1500 kPa water
(percentages) of 1.25 or less; or
(2) A ratio of CEC (at pH near 8) to 1500 kPa water of
>1.5 and more exchange acidity than the sum of bases
plus KCl-extractable aluminum;
g. When neither irrigated nor fallowed to store moisture:
(1) If the soil temperature regime is mesic or thermic,
are dry less than sixth-tenths of the time in half or more
years in some part of the moisture control section (not
necessarily the same part) during the period when the soil
temperature, at a depth 50 cm below the surface, exceeds
5°C;
(2) If the soil temperature regime is hyperthermic or
isomesic or warmer, are moist in some or all parts of the
soil moisture control section for 90 consecutive days or
more during a period when the soil temperature at a
depth 50 cm below the soil surface is higher than 8°C.
Andic Ustochrepts are like Typic Ustochrepts except for *f* or
for *f* and *e.*
Aridic Ustochrepts are like Typic Ustochrepts except for *g.*
Fluventic Ustochrepts are like Typic Ustochrepts except for
a.
Lithic Ustochrepts are like Typic Ustochrepts except for *b,*
with or without *e* or *g,* or both.
Udertic Ustochrepts are like Typic Ustochrepts except for *c*
and *e,* with or without *a* or *d,* or both.
Udic Ustochrepts are like Typic Ustochrepts except for *e.*
Vertic Ustochrepts are like Typic Ustochrepts except for *c,*
with or without *a* or *d,* or both.

Xerochrepts

Distinctions between Typic Xerochrepts and other subgroups

Typic Xerochrepts are the Xerochrepts that
a. Do not have mottles that have chroma of 2 or less within
75 cm of the soil surface;
b. Have base saturation (by NH_4OAc) of 60 percent or more
in some part of the soil between depths of 25 and 75 cm
below the soil surface;
c. Have a content of organic carbon that decreases regularly
with depth, and unless a lithic or a paralithic contact occurs
at a shallower depth, reaches a level of 0.2 percent or less
within 1.25 m of the soil surface; or have slopes >25
percent;
d. Do not have a lithic contact within 50 cm of the soil
surface;
e. Do not have the following combination of characteristics:
(1) Cracks at some period in most years that are 1 cm or
more wide at a depth of 50 cm, that are at least 30 cm
long in some part, and that extend upward to the soil
surface or to the base of an Ap horizon;

(2) A coefficient of linear extensibility (COLE) of 0.05 or more in a horizon or horizons at least 50 cm thick and a potential linear extensibility of 6 cm or more in the upper 1.5 m of the soil or in the whole soil if a lithic or paralithic contact is deeper than 50 cm but shallower than 1.5 m; and

(3) More than 35 percent clay in horizons that total >50 cm in thickness; or

f. Do not have a calcic horizon or soft powdery lime within a depth of 1.5 m if the weighted average particle-size class from depths of 25 cm to 1 m is sandy or to a lithic or paralithic contact if one is shallower than 1 m, or within a depth of 1.1 m if the weighted average particle-size class is loamy, or within a depth of 90 cm if it is clayey;

g. Do not have a layer in the upper 75 cm that has texture finer than loamy fine sand, that is a much as 18 cm thick, that has bulk density (at 33 kPa water tension) of 0.95 g per cubic centimeter or less in the fine-earth fraction, and that has either of the following:

(1) A ratio of measured clay to 1500 kPa water (percentages) of 1.25 or less; or

(2) A ratio of CEC (at pH near 8) to 1500 kPa water of >1.5 and more exchange acidity than the sum of bases plus KCl-extractable aluminum.

h. Do not have a petrocalcic horizon within a depth of 100 cm of the soil surface[11].

i. Do not have a gypsic horizon within a depth of 100 cm of the soil surface[12].

Andic Xerochrepts are like Typic Xerochrepts except for *g* or for *b* and *g*.

Aquic Xerochrepts are like Typic Xerochrepts except for *a*.

Aquic Dystric Xerochrepts are like Typic Xerochrepts except for *a* and *b*.

Calcixerollic Xerochrepts are like Typic Xerochrepts except for *f*.

Dystric Xerochrepts are like Typic Xerochrepts except for *b*.

Dystric Fluventic Xerochrepts are like Typic Xerochrepts except for *b* and *c*.

Dystric Lithic Xerochrepts are like Typic Xerochrepts except for *b* and *d*.

Fluventic Xerochrepts are like Typic Xerochrepts except for *c*.

Gypsic Xerochrepts are like Typic Xerochrepts except for *i*, with or without *f*.

Lithic Xerochrepts are like Typic Xerochrepts except for *d*.

Lithic Ruptic-Xerorthentic Xerochrepts are like Typic Xerochrepts except for *d*, and have an intermittent cambic horizon.

Petrocalcic Xerochrepts are like Typic Xerochrepts except for *h*, with or without *f* or *i* or both.

Ruptic-Lithic Xerochrepts are like Typic Xerochrepts except for *d* and have a lithic contact in some part but less than half of each pedon.

Vertic Xerochrepts are like Typic Xerochrepts except for *e*, with or without *a* or *c*, or both.

---

[11]Personal communications with Dr. R. Tavernier and Ph.D. thesis by M. Ilaiwi, Univ. of Ghent.

[12]See footnote 11 above.

Plaggepts

Plaggepts are the soils that have a plaggen epipedon that is composed of crystalline rather than pyroclastic materials. This suborder includes all freely drained soils that have a plaggen epipedon except a few Andepts.

Tropepts

Key to great groups

IDA. Tropepts that have base saturation of <50 percent (by NH_4OAc) in some subhorizon between depths of 25 cm and 1 m and have 12 kg or more organic carbon, exclusive of surface litter, per square meter in the soil to a depth of 1 m, or to a lithic, paralithic, or petroferric contact if one is shallower than 1 m, and do not have a sombric horizon.

Humitropepts, p. 165

IDB. Other Tropepts that have a sombric horizon.

Sombritropepts, p. 166

IDC. Other Tropepts that have an ustic moisture regime or have soft powdery lime within 1.5 m of the soil surface and have base saturation (by NH_4OAc) of 50 percent or more in all subhorizons between depths of 25 cm and 1 m, or between 25 cm and a lithic, paralithic or petroferric contact if one is shallower than 1 m.

Ustropepts, p. 166

IDD. Other Tropepts that have base saturation (by NH_4OAc) of 50 percent or more in all subhorizons between depths of 25 cm and 1 m, or between 25 cm and a lithic or paralithic contact if one is shallower than 1 m.

Eutropepts, p. 164

IDE. Other Tropepts.

Dystropepts, p. 163

Dystropepts

Distinctions between Typic Dystropepts and other subgroups

Typic Dystropepts are the Dystropepts that
a. Do not have mottles that have chroma of 2 or less within 1 m of the soil surface if the mottled horizon is saturated with water at some time of year or the soil has artificial drainage.
b. Do not have a layer in the upper 75 cm that has texture finer than loamy fine sand, that is as much as 18 cm thick, that has bulk density (at 33 kPa water tension) of 0.95 g per cubic centimeter or less in the fine-earth fraction, and that has either of the following:
    (1) A ratio of measured clay to 1500 kPa water (percentages) of 1.25 or less; or
    (2) A ratio of CEC (at pH near 8) to 1500 kPa water of >1.5 and more exchange acidity than the sum of bases plus KCl-extractable aluminum;
c. Have a cambic horizon;

d. Have a content of organic carbon[13] that decreases regularly with depth and, unless a lithic or paralithic contact is present at a shallower depth, reaches a level of 0.2 percent or less within 1.25 m of the soil surface, or have slopes >25 percent;

e. Do not have a lithic or a petroferric contact within 50 cm of the soil surface;

f. Have a CEC (by 1 M NH$_4$OAc pH 7) of 24 or more cmol(+) kg^{-1} clay[14] in the major part of the soil below a depth of 25 cm but above 100 cm or a lithic or paralithic contact if one is shallower than 100 cm.

g. Have a udic moisture regime; and

h. Do not have the following combination of characteristics:

(1) Cracks at some period in most years that are 1 cm or more wide at a depth of 50 cm, that are at least 30 cm long in some part, and that extend upward to the soil surface or to the base of an Ap horizon;

(2) A coefficient of linear extensibility (COLE) of 0.09 or more if the soil moisture regime is udic, or 0.07 or more if it is ustic, in a horizon or horizons at least 50 cm thick, and a potential linear extensibility of 6 cm or more in the upper 1 m or 1.25 m, respectively, of the soil or in the whole soil if a lithic or paralithic contact is deeper than 50 cm but shallower than 1 m or 1.25 m;

(3) More than 35 percent clay in horizons that total >50 cm in thickness.

Aquic Dystropepts are like Typic Dystropepts except for *a*.

Fluventic Dystropepts are like Typic Dystropepts except for *d*.

Lithic Dystropepts are like Typic Dystropepts except for *e* and have a lithic contact within a depth of 50 cm.

Oxic Dystropepts are like Typic Dystropepts except for *f*.

Ustic Dystropepts are like Typic Dystropepts except for *g*, and they have an ustic moisture regime.

Ustoxic Dystropepts are like Typic Dystropepts except for *f* and *g*, and they have an ustic moisture regime.

Vertic Dystropepts are like Typic Dystropepts except for *h*, with or without any or all of *a*, *c*, *d*, and *g*.

Eutropepts

Distinctions between Typic Eutropepts and other subgroups

Typic Eutropepts are the Eutropepts that

a. Do not have a layer in the upper 75 cm that has texture finer than loamy fine sand, that is as much as 18 cm thick, that has bulk density (at 33 kPa water tension) of 0.95 g per cubic centimeter or less in the fine-earth fraction, and that has either of the following:

(1) A ratio of measured clay to 1500 kPa water (percentages) of 1.25 or less; or

---

[13] The carbon should be of Holocene age. It is not the intent to include fossil carbon from bedrock.

[14] Some cambic horizons that have properties that approach those of an oxic horizon do not disperse well. If the ratio of the percentage of water retained at tension of 1500 kPa to the percentage of measured clay is 0.6 or more, the percentage of clay is determined by the higher value of (1) the measured percentage of clay or (2) 2.5 times the percentage of water retained at tension of 1500 kPa.

(2) A ratio of CEC (at pH near 8) to 1500 kPa water of
>1.5 and more exchange acidity than the sum of bases
plus KCl-extractable aluminum;
b. Do not have mottles that have chroma of 2 or less within
1 m of the soil surface if the mottled horizon is saturated
with water at some time during the year or if there is
artificial drainage.
c. Have a content of organic carbon[15] that decreases
regularly with depth and, unless a lithic or a paralithic
contact is present at a shallower depth, reaches a level of 0.2
percent organic carbon or less within 1.25 m of the soil
surface; or have slopes >25 percent;
d. Do not have a lithic contact within 50 cm of the soil
surface;
e. Do not have the following combination of characteristics:
(1) Cracks at some period in most years that are 1 cm or
more wide at a depth of 50 cm, that are at least 30 cm
long in some part and that extend upward to the soil
surface or to the base of an Ap horizon;
(2) A coefficient of linear extensibility (COLE) of 0.09
or more in a horizon or horizons at least 50 cm thick and
a potential linear extensibility of 6 cm or more in the
upper 1 m of the soil, or in the whole soil if a lithic or
paralithic contact is deeper than 50 cm but shallower than
1 m; and
(3) More than 35 percent clay in horizons that total >50
cm in thickness; or
f. Have a cambic horizon.
Andic Eutropepts are like Typic Eutropepts except for *a*.
Aquic Eutropepts are like Typic Eutropepts except for *b*.
Fluvaquentic Eutropepts are like Typic Eutropepts except for
b and *c*.
Fluventic Eutropepts are like Typic Eutropepts except for *c*.
Lithic Eutropepts are like Typic Eutropepts except for *d*.
Vertic Eutropepts are like Typic Eutropepts except for *e*,
with or without any or all of *b*, *c*, or *f*.

Humitropepts

Distinctions between Typic Humitropepts and other
subgroups

Typic Humitropepts are the Humitropepts that
a. Do not have a layer in the upper 75 cm that has texture
finer than loamy fine sand, that is as much as 18 cm thick,
that has bulk density (at 33 kPa water tension) of 0.95 g per
cubic centimeter or less in the fine-earth fraction, and that
has either
(1) a ratio of measured clay to 1500 kPa water
(percentages) of 1.25 or less or
(2) a ratio of CEC (at pH near 8) to 1500 kPa water of
>1.5 and more exchange acidity than the sum of bases
plus KCl-extractable aluminum;
b. Do not have mottles that have chroma of 2 or less within
1 m of the soil surface if the mottled horizon is saturated
with water at some time of year or if there is artificial
drainage;

---

[15]The carbon should be of Holocene age. It is not the intent to include fossil
carbon from bedrock.

c. Do not have a lithic contact within 50 cm of the soil surface;

d. Have a CEC (by NH_4OAc pH 7) of 24 or more cmol(+) kg^{-1} clay[16] in the major part of the soil below a depth of 25 cm but above 100 cm or a lithic or paralithic contact if one is shallower than 100 cm.

e. Have a content of organic carbon that decreases regularly with depth to the base of the cambic horizon, or have slopes >25 percent; and

f. Have a udic moisture regime.

Andic Humitropepts are like Typic Humitropepts except for *a* or for *a* and *e*.

Andic Ustic Humitropepts are like Typic Humitropepts except for *a* and *f*, with or without *e*, and they have an ustic moisture regime.

Fluventic Humitropepts are like Typic Humitropepts except for *e*.

Lithic Humitropepts are like Typic Humitropepts except for *c*.

Oxic Humitropepts are like Typic Humitropepts except for *d*.

Ustic Humitropepts are like Typic Humitropepts except for *f*, and they have an ustic moisture regime.

Ustoxic Humitropepts are like Typic Humitropepts except for *d* and *f*, and they have an ustic moisture regime.

Sombritropepts

These soils are the dark, humus-rich Tropepts of perhumid, cool, hilly or mountainous regions. They have a sombric horizon in or below a cambic horizon. Most of them have an umbric epipedon, a perudic soil moisture regime, and an isomesic temperature regime. They are not known to occur in the United States and their classification has not been developed.

Ustropepts

Distinctions between Typic Ustropepts and other subgroups

Typic Ustropepts are the Ustropepts that

a. Do not have mottles that have chroma of 2 or less within 1 m of the soil surface if the mottled horizon is saturated with water at some time of the year or if there is artificial drainage;

b. Have a content of organic carbon that decreases regularly with depth and, unless a lithic or a paralithic contact occurs at a shallower depth, reaches a level of 0.2 percent organic carbon or less within 1.25 m of the soil surface; or have slopes >25 percent;

c. Have a cambic horizon;

d. Do not have a lithic contact within 50 cm of the soil surface;

---

[16]Some cambic horizons that have properties that approach those of an oxic horizon do not disperse well. If the ratio of the percentage of water retained at tension of 1500 kPa to the percentage of measured clay is 0.6 or more, the percentage of clay is determined by the higher value of (1) the measured percentage of clay or (2) 2.5 times the percentage of water retained at tension of 1500 kPa.

e. Have a CEC (by NH_4OAc pH7) of 24 or more cmol(+) kg^{-1} clay[17] in the major part of the soil below a depth of 25 cm but above 100 cm or a lithic or paralithic contact if one is shallower than 100 cm; and

f. Do not have the following combination of characteristics:

(1) Cracks at some period in most years that are 1 cm or more wide at a depth of 50 cm, that are at least 30 cm long in some part, and that extend upward to the soil surface or to the base of an Ap horizon;

(2) A coefficient of linear extensibility (COLE) of 0.07 or more in a horizon or horizons at least 50 cm thick and a potential linear extensibility of 6 cm or more in the upper 1.25 m of the soil or in the whole soil if a lithic or paralithic contact is deeper than 50 cm but shallower than 1.25 m; and

(3) More than 35 percent clay in horizons that total >50 cm in thickness;

Fluventic Ustropepts are like Typic Ustropepts except for *b*, with or without *c* or *e*, or both, and the organic-carbon content is ≥ 0.5 percent at a depth of 1.25 m of the cambic horizon or of the whole soil if there is no cambic horizon.

Lithic Ustropepts are like Typic Ustropepts except for *d*.

Oxic Ustropepts are like Typic Ustropepts except for *e* or for *e* and *b*, and the organic-carbon content is <0.5 percent at a depth 1.25 m below the soil surface.

Vertic Ustropepts are like Typic Ustropepts except for *f*(1), with or without any or all of *a*, *b*, or *c*.

Umbrepts

Key to great groups

IFA. Umbrepts that have a fragipan.

Fragiumbrepts, p. 168

IFB. Other Umbrepts that have a cryic or pergelic temperature regime.

Cryumbrepts, p. 167

IFC. Other Umbrepts that have a xeric moisture regime.

Xerumbrepts, p. 169

IFD. Other Umbrepts.

Haplumbrepts, p. 169

Cryumbrepts

Distinctions between Typic Cryumbrepts and other subgroups

Typic Cryumbrepts are the Cryumbrepts that

a. Do not have a layer in the upper 75 cm that has texture finer than loamy fine sand, that is as much as 18 cm thick, that has bulk density (at 33 kPa water tension) of 0.95 g per

---

[17]Some cambic horizons that have properties that approach those of an oxic horizon do not disperse well. If the ratio of the percentage of water retained at tension of 1500 kPa to the percentage of measured clay is 0.6 or more, the percentage of clay is determined by the higher value of (1) the measured percentage of clay or (2) 2.5 times the percentage of water retained at tension of 1500 kPa.

cubic centimeter or less in the fine-earth fraction, and that has either
(1) a ratio of measured clay to 1500 kPa water (percentages) of 1.25 or less or
(2) a ratio of CEC (at pH near 8) to 1500 kPa water of >1.5 and more exchange acidity than the sum of bases plus KCl-extractable aluminum;
b. Have a cambic horizon;
c. Do not have a lithic contact within 50 cm of the surface;
d. Have a mean annual soil temperature higher than 0°C;
e. Do not have mottles that have chroma of 2 or less within 75 cm of the soil surface if the mottled horizon is saturated with water at some time of the year when its temperature is ≥ 5°C or if there is artificial drainage; and
f. Have an umbric epipedon that is continuous in each pedon.
Andic Cryumbrepts are like Typic Cryumbrepts except for *a*.
Aquic Cryumbrepts are like Typic Cryumbrepts except for *e* or for *b* and *e*.
Entic Cryumbrepts are like Typic Cryumbrepts except for *b*.
Lithic Cryumbrepts are like Typic Cryumbrepts except for *c*, with or without *b* or *d*, or both.
Lithic Ruptic-Entic Cryumbrepts are like Typic Cryumbrepts except for *c* and *f*, with or without *b* or *d*, or both.
Pergelic Cryumbrepts are like Typic Cryumbrepts except for *d*, with or without *a* or *b* or both.
Ruptic-Lithic Cryumbrepts are like Typic Cryumbrepts except for *c*, with or without *b* or *d*, or both, and they have a lithic contact within 50 cm of the surface in only part of each pedon.

Fragiumbrepts

Distinctions between Typic Fragiumbrepts and other subgroups

The definitions that follow are incomplete because there are few of these soils in the United States.

Typic Fragiumbrepts are the Fragiumbrepts that
a. Do not have a layer in the upper 75 cm that has texture finer than loamy fine sand, that is as much as 18 cm thick, that has bulk density (at 33 kPa water tension) of 0.95 g per cubic centimeter or less in the fine-earth fraction, and that has either
(1) a ratio of measured clay to 1500 kPa water (percentages) of 1.25 or less or
(2) a ratio of CEC (at pH near 8) to 1500 kPa water of >1.5 and more exchange acidity than the sum of bases plus KCl-extractable aluminum; and
b. Do not have mottles that have chroma of 2 or less within 50 cm of the soil surface.
Andic Fragiumbrepts are like Typic Fragiumbrepts except for *a*.
Aquic Fragiumbrepts are like Typic Fragiumbrepts except for *b*.

Haplumbrepts

Distinctions between Typic Haplumbrepts and other subgroups

Typic Haplumbrepts are the Haplumbrepts that
a. Do not have a layer in the upper 75 cm that has texture finer than loamy fine sand, that is as much as 18 cm thick, that has bulk density (at 33 kPa water tension) of 0.95 g per cubic centimeter or less in the fine-earth fraction, and that has either
    (1) a ratio of measured clay to 1500 kPa water (percentages) of 1.25 or less or
    (2) a ratio of CEC (at pH near 8) to 1500 kPa water of >1.5 and more exchange acidity than the sum of bases plus KCl-extractable aluminum;
b. Do not have mottles that have chroma of 2 or less within 50 cm of the soil surface if the mottled horizon is saturated with water at some time of the year when its temperature is >5°C or if there is artificial drainage;
c. Have a cambic horizon;
d. Have a content of organic carbon[18] that decreases regularly with depth or have slopes >25 percent;
e. Do not have a lithic contact within 50 cm of the soil surface;
f. Have an umbric or a mollic epipedon that is <50 cm thick; and
g. Have texture finer than loamy fine sand within a depth of 50 cm.
Andic Haplumbrepts are like Typic Haplumbrepts except for *a* or for *a* and *f.*
Andaquic Haplumbrepts are like Typic Haplumbrepts except for *a* and *b* or *a*, *b*, and *f.*
Cumulic Haplumbrepts are like Typic Haplumbrepts except for *f* and *d*, with or without *b* or *c*, or both.
Entic Haplumbrepts are like Typic Haplumbrepts except for *c.*
Fluventic Haplumbrepts are like Typic Haplumbrepts except for *d.*
Lithic Haplumbrepts are like Typic Haplumbrepts except for *e* or for *c* and *e.*
Pachic Haplumbrepts are like Typic Haplumbrepts except for *f*, with or without *b* or *c*, or both.
Quartzipsammentic Haplumbrepts are like Typic Haplumbrepts except for *c* and *g*, with or without *f.* They have sandy texture to a depth of 1 m or more, and have in the 0.02 to 2mm fraction more than 90 percent silica minerals (quartz, chalcedony or opal) or other extremely durable minerals that are resistant to weathering.

Xerumbrepts

Distinctions between Typic Xerumbrepts and other subgroups

Typic Xerumbrepts are the Xerumbrepts that
a. Do not have mottles that have chroma of 2 or less within 75 cm of the soil surface if the mottled horizon is saturated

---

[18]The carbon should be of Holocene age. It is not the intent to include fossil carbon from bedrock.

with water at some time of the year when its temperature is >5°C or if there is artificial drainage;

b. Have an umbric or mollic epipedon that is <50 cm thick;

c. Have a cambic horizon;

d. Have a content of organic carbon that decreases regularly with depth or have slopes >25 percent;

e. Do not have a lithic contact within 50 cm of the soil surface; and

f. Do not have a layer in the upper 75 cm that has texture finer than loamy fine sand, that is as much as 18 cm thick, that has bulk density (at 33 kPa water tension) of 0.95 g per cubic centimeter or less in the fine-earth fraction, and that has either

> **(1)** a ratio of measured clay to 1500 kPa water (percentages) of 1.25 or less or

> **(2)** a ratio of CEC (at pH near 8) to 1500 kPa water of >1.5 and more exchange acidity than the sum of bases plus KCl-extractable aluminum.

Andic Xerumbrepts are like Typic Xerumbrepts except for *f*.

Entic Xerumbrepts are like Typic Xerumbrepts except for *c*.

Fluventic Xerumbrepts are like Typic Xerumbrepts except for *d*.

Lithic Xerumbrepts are like Typic Xerumbrepts except for *e*, with or without *c*.

Pachic Xerumbrepts are like Typic Xerumbrepts except for *b*, with or without *a* or *c*, or both.

Chapter 10

Mollisols

Key To Suborders

GA. Mollisols that have all the following:
  1. An albic horizon that lies immediately under the mollic epipedon or that separates horizons that together meet all the requirements of a mollic epipedon;
  2. An argillic or a natric horizon; and
  3. Chroma of 2 or less in the albic horizon or characteristics associated with wetness in the albic, argillic, or natric horizon, namely mottles or iron-manganese concretions larger than 2 mm or both.

Albolls, p. 172

GB. Other Mollisols that either have an aquic moisture regime or are artificially drained, and that have one or more of the following characteristics associated with wetness:
  1. A histic epipedon overlying the mollic epipedon;
  2. An SAR ≥ 13 (or sodium saturation of ≥ 15 percent) in the upper part of the mollic epipedon and decreasing SAR (or sodium saturation) with increasing depth below 50 cm;
  3. One of the following combinations of colors, moist;
    a. If the lower part of the mollic epipedon[1] has chroma of 1 or less, there are either
      (1) Distinct or prominent mottles in the lower part of the mollic epipedon; or
      (2) A color value, moist, of 4 or more immediately below the mollic epipedon, or within 75 cm of the surface if a calcic horizon intervenes, and one of the following:
        (a) If the hue is 10YR or redder and there are mottles, chroma is less than 1.5 on ped surfaces or in the matrix; if there are no mottles, chroma is less than 1;
        (b) If the hue is nearest 2.5Y and there are distinct or prominent mottles, chroma is 2 or less on ped surfaces or in the matrix; if there are no mottles, chroma is 1 or less;
        (c) If the nearest hue is 5Y or yellower and there are distinct or prominent mottles, chroma is 3 or less on ped surfaces or in the matrix; if there are no mottles, chroma is 1 or less;
        (d) The hue is bluer than 10Y or the color is neutral; or
        (e) The color results from uncoated mineral grains; or
    b. If the lower part of the mollic epipedon has chroma of more than 1 but not more than 2, there are either
      (1) Distinct or prominent mottles in the lower mollic epipedon; or
      (2) Base colors immediately below the mollic epipedon that have one or more of the following properties:

MOL

---

[1]If the mollic epipedon extends to a lithic contact within 30 cm of the surface, the requirement for mottles is waived.

(a) Value of 4 and chroma of 2 and also some mottles that have value of 4 or more and chroma less than 2;
(b) Value of 5 or more and chroma of 2 or less and also mottles that have higher chroma; or
(c) Value of 4 and chroma <2; or
4. A calcic or petrocalcic horizon that has its upper boundary within 40 cm of the surface.

Aquolls, p. 173

GC. Other Mollisols that have all the following characteristics:
1. Have a mollic epipedon that is not more than 50 cm thick;
2. Do not have an argillic horizon;
3. Do not have a calcic horizon;
4. The soil materials in or immediately below any mollic epipedon, including coarse fragments less than 7.5 cm in diameter, have a $CaCO_3$ equivalent of 40 percent or more; and
5. Have a udic moisture regime or a cryic temperature regime;

Rendolls, p. 183

GD. Other Mollisols that have a xeric moisture regime or an aridic moisture regime bordering on xeric but do not have a cryic temperature regime.

Xerolls, p. 195

GE. Other Mollisols that have a frigid, cryic, or pergelic temperature regime.

Borolls, p. 176

GF. Other Mollisols that have an ustic or an aridic moisture regime that borders on ustic.

Ustolls, p. 187

GG. Other Mollisols.

Udolls, p. 184

Albolls

Key to great groups

GAA. Albolls that have a natric horizon.

Natralbolls, p. 173

GAB. Other Albolls.

Argialbolls, p. 172

Argialbolls

Distinctions between Typic Argialbolls and other subgroups

Typic Argialbolls are the Argialbolls that
a. Do not have a layer in the upper 75 cm that has a texture finer than loamy fine sand, that is as much as 18 cm thick, that has a bulk density (at 33 kPa water tension) of 0.95 g per cubic centimeter or less in the fine-earth fraction, and that has either

(1) a ratio of measured clay to 1500 kPa water
(percentages)of 1.25 or less or
(2) a ratio of CEC (at pH near 8) to 1500 kPa water of
>1.5 and more exchange acidity than the sum of bases
plus KCl-extractable aluminum;
b. Have an abrupt textural change from the albic to the
argillic horizon;
c. When not irrigated, are not dry in all parts of the
moisture control section for as long as 45 consecutive days
during the 120 days following the summer solstice in more
than 6 out of 10 years.
Argiaquic Argialbolls are like Typic Argialbolls except for *b.*
Argiaquic Xeric Argialbolls are like Typic Argialbolls except
for *b* and *c.*
Xeric Argialbolls are like Typic Argialbolls except for *c.*

Natralbolls

Natralbolls are the Albolls that have a natric horizon.

Aquolls

Key to great groups

GBA. Aquolls that have a cryic or pergelic temperature
regime.

Cryaquolls, p. 174

GBB. Other Aquolls that have a duripan that has its upper
boundary within 1 m of the surface.

Duraquolls, p. 175

GBC. Other Aquolls that have a natric horizon.

Natraquolls, p. 176

GBD. Other Aquolls that have a calcic or gypsic horizon
that has its upper boundary within 40 cm of the surface and
do not have an argillic horizon unless it is a buried horizon.

Calciaquolls, p. 174

GBE. Other Aquolls that have an argillic horizon.

Argiaquolls, p. 173

GBF. Other Aquolls.

Haplaquolls, p. 175

Argiaquolls

Distinctions between Typic Argiaquolls and other subgroups

Typic Argiaquolls are the Argiaquolls that
a. Do not have an argillic horizon that has an increase in
clay content of 20 percent (absolute) or more within a
vertical distance of 7.5 cm from the upper boundary;
b. Have a texture finer than loamy fine sand in some
subhorizon within 50 cm of the surface; and
c. Do not have the following combination of characteristics:
   (1) Cracks at some period in most years that are 1 cm or
   more wide at a depth of 50 cm, that are at least 30 cm
   long in some part, and that extend upward to the surface
   of the soil or to the base of the Ap horizon,

(2) A coefficient of linear extensibility (COLE) of 0.09 or more in a horizon or horizons at least 50 cm thick and a potential linear extensibility of 6 cm or more in the soil to a depth of 1 m or in the whole soil if a lithic or paralithic contact is deeper than 50 cm but shallower than 1 m, and

(3) More than 35 percent clay in horizons that total >50 cm in thickness.

Abruptic Argiaquolls are like Typic Argiaquolls except for *a*.
Arenic Argiaquolls are like Typic Argiaquolls except for *b* and have a sandy epipedon between 50 cm and 1 m thick.
Grossarenic Argiaquolls are like Typic Argiaquolls except for *b* and have a sandy epipedon >1 m thick.
Vertic Argiaquolls are like Typic Argiaquolls except *c*.

Calciaquolls

Distinctions between Typic Calciaquolls and other subgroups

Typic Calciaquolls are the Calciaquolls that
a. Do not have color that has dominant chroma of 3 or more in the matrix or on the ped surfaces in any subhorizon within 75 cm of the surface and do have one or more of the following colors immediately below the mollic epipedon:

(1) If the nearest hue is 2.5Y or yellower and there are distinct or prominent mottles, the chroma, moist, is 2 or less; if there are no mottles, the chroma, moist, is 1 or less; or

(2) If the nearest hue is 10YR or redder and there are distinct or prominent mottles, the chroma is nearer to 1 than to 2; if there are no mottles, the chroma is 1 or less; and

b. Do not have a petrocalcic horizon that has its upper boundary within 1 m of the surface.

Aeric Calciaquolls are like Typic Calciaquolls except for *a*.
Petrocalcic Calciaquolls are like Typic Calciaquolls except for *b*.

Cryaquolls

Distinctions between Typic Cryaquolls and other subgroups

Typic Cryaquolls are the Cryaquolls that
a. Do not have a layer in the upper 75 cm that has texture finer than loamy sand, that is as much as 18 cm thick, that has a bulk density (at 33 kPa water tension) in the fine-earth fraction of 0.95 g per cubic centimeter or less, and that has either

(1) a ratio of measured clay to 1500 kPa water (percentages) of 1.25 or less or

(2) a ratio of CEC (at pH near 8) to 1500 kPa water of >1.5 and more exchange acidity than the sum of bases plus KCl-extractable aluminum;

b. Do not have an argillic horizon;
c. Do not have a calcic horizon within or immediately underlying the mollic epipedon;
d. Have a mollic epipedon that is <50 cm thick;
e. Do not have a histic epipedon;
f. Have a mean annual soil temperature higher than $0^{\circ}C$; and

g. Do not have a buried Histosol that has its upper boundary within a depth of 1 m.

Argic Cryaquolls are like Typic Cryaquolls except for *b*, for *b* and *c*, or for *b*, *c*, and *d*.

Calcic Cryaquolls are like Typic Cryaquolls except for *c*.

Cumulic Cryaquolls are like Typic Cryaquolls except for *d*.

Histic Cryaquolls are like Typic Cryaquolls except for *e*.

Pergelic Cryaquolls are like Typic Cryaquolls except for *f* or for *e* and *f*.

Thapto-Histic Cryaquolls are like Typic Cryaquolls except for *g* or for *g* and *d*.

Duraquolls

Distinctions between Typic Duraquolls and other subgroups

Typic Duraquolls are the Duraquolls that
a. Do not have an argillic horizon; and
b. Do not have a natric horizon.

Argic Duraquolls are like Typic Duraquolls except for *a*.

Natric Duraquolls are like Typic Duraquolls except for *b*.

Haplaquolls

Distinctions between Typic Haplaquolls and other subgroups

Typic Haplaquolls are the Haplaquolls that
a. Do not have a layer in the upper 75 cm that has a texture finer than loamy fine sand, that is as much as 18 cm thick, that has a bulk density (at 33 kPa water tension) of 0.95 g per cubic centimeter or less in the fine-earth fraction, and that has either of the following:
   (1) A ratio of measured clay to 1500 kPa water (percentages) of 1.25 or less; or
   (2) A ratio of CEC (at pH near 8) to 1500 kPa water of >1.5 and more exchange acidity than the sum of bases plus KCl-extractable aluminum;
b. Do not have a buried Histosol that has its upper boundary within 1 m of the soil surface;
c. Have a mollic epipedon that is <60 cm thick;
d. Do not have a horizon 15 cm or more thick that is within 1 m of the surface and that contains at least 20 percent (by volume) of durinodes or is brittle and has firm consistence when moist;
e. Have a content of organic carbon that decreases regularly with increasing depth and reaches a level of 0.3 percent carbon or less in some subhorizon within 1.25 m of the soil surface, or the slope is >25 percent;
f. Do not have a histic epipedon;
g. Do not have a lithic contact within 50 cm of the surface; and
h. Do not have the following combination of characteristics:
   (1) Cracks at some period in most years that are 1 cm or more wide at a depth of 50 cm, that are at least 30 cm long in some part, and that extend upward to the surface or to the base of an Ap horizon; and
   (2) A coefficient of linear extensibility (COLE) of 0.09 or more in a horizon or horizons at least 50 cm thick and a potential linear extensibility of 6 cm or more in the upper 1 m of the soil or in the whole soil if a lithic or

paralithic contact is deeper than 50 cm but shallower than 1 m; and
(3) More than 35 percent clay in horizons that total >50 cm in thickness.
Andaqueptic Haplaquolls are like Typic Haplaquolls except for *a* or for *a* and *c*.
Cumulic Haplaquolls are like Typic Haplaquolls except for *c* or for *c* and *e*.
Duric Haplaquolls are like Typic Haplaquolls except for *d*.
Fluvaquentic Haplaquolls are like Typic Haplaquolls except for *e*.
Histic Haplaquolls are like Typic Haplaquolls except for *f*.
Lithic Haplaquolls are like Typic Haplaquolls except for *g*.
Vertic Haplaquolls are like Typic Haplaquolls except for *h*, with or without *c* or *e*, or both.

Natraquolls

Natraquolls are the Aquolls that
1. Have a natric horizon; and
2. Do not have a duripan that has its upper boundary within 1 m of the soil surface.

Borolls

Key to great groups

GEA. Borolls that have an argillic horizon that has its upper boundary deeper than 60 cm below the mineral soil surface[2] and that have texture finer than loamy fine sand in all subhorizons above the argillic horizon.

Paleborolls, p. 182

GEB. Other Borolls that have a cryic or pergelic temperature regime.

Cryoborolls, p. 179

GEC. Other Borolls that have a natric horizon but do not have a cambic horizon that is above the natric horizon and separated from it by an albic horizon.

Natriborolls, p. 182

GED. Other Borolls that have an argillic horizon but do not have a cambic horizon that is above the argillic horizon and separated from it by an albic horizon.

Argiborolls, p. 177

GEE. Other Borolls that have a mollic epipedon that, below any Ap horizon, is 50 percent or more by volume wormholes, wormcasts, or filled animal burrows and that either rests on a lithic contact or has a transition to the underlying horizon in which 25 percent or more of the material is discrete wormholes, wormcasts, or animal burrows filled with material from the mollic epipedon and the underlying horizon.

Vermiborolls, p. 183

---

[2]If there is a surface mantle that has >60 percent vitric volcanic ash, cinders, or other vitric pyroclastic materials, the depth to the argillic horizon is measured from the base of this mantle rather than from the mineral soil surface.

GEF. Other Borolls that have a calcic or petrocalcic horizon whose upper boundary is within 1 m of the soil surface and that are calcareous in all parts of all horizons above the calcic or petrocalcic horizon, after the upper soil to a depth of 18 cm has been mixed, unless the texture is coarser than loamy very fine sand.

Calciborolls, p. 178

GEG. Other Borolls.

Haploborolls, p. 180

Argiborolls

Distinctions between Typic Argiborolls and other subgroups

Typic Argiborolls are the Argiborolls that
a. Do not have an argillic horizon that has an increase in clay content of 20 percent (absolute) or more within a vertical distance of 7.5 cm from the upper boundary;
b. Do not have either
 (1) An albic horizon that lies immediately under the mollic epipedon; or
 (2) Tonguing or interfingering of albic materials in the upper part of the argillic horizon, or skeletans of clean silt and sand covering more than half the ped faces in the upper 5 cm or more of the argillic horizon;
c. Do not have a layer in the upper 75 cm that has a texture finer than loamy fine sand, that is as much as 18 cm thick, that has a bulk density (at 33 kPa water tension) of 0.95 g per cubic centimeter or less in the fine-earth fraction, and that has either of the following:
 (1) A ratio of measured clay to 1500 kPa water (percentages) of 1.25 or less; or
 (2) A ratio of CEC (at pH near 8) to 1500 kPa water of >1.5 and more exchange acidity than the sum of bases plus KCl-extractable aluminum;
d. Do not have mottles that have chroma of 2 or less within 1 m of the surface or, if undrained, are not continuously saturated with water for as long as 90 days within 1 m of the surface;
e. Do not have a lithic contact within 50 cm of the surface;
f. Have
 (1) Either or both
  (a) A color value, dry, of less than 4.5 in the upper 18 cm of the mollic epipedon, after mixing, or in any Ap horizon that is >18 cm thick; or
  (b) A moisture control section that is dry in some part less than six-tenths of the time that the soil temperature at a depth of 50 cm is above 5°C in most years; and
 (2) A chroma (rubbed), moist, of 1.5 or more in the upper 18 cm of the mollic epipedon after mixing, or in any Ap horizon that is >18 cm thick and the soil is dry in all parts of the moisture control section at some time in most years;
g. Have a mollic epipedon that is <40 cm thick, or its texture is loamy fine sand or coarser; and
h. Do not have the following combination of characteristics:
 (1) Cracks at some period in most years that are 1 cm or more wide at a depth of 50 cm, that are at least 30 cm

long in some part, and that extend upward to the surface
or to the base of an Ap horizon;
(2) A potential linear extensibility of 6 cm or more in
the upper 1 m of the soil or in the whole soil if a lithic
or paralithic contact is deeper than 50 cm but shallower
than 1 m; and
(3) More than 35 percent clay in horizons that total >50
cm in thickness.
Abruptic Argiborolls are like Typic Argiborolls except for *a*,
with or without *b(1)* or *h*, or both.
Abruptic Aridic Argiborolls are like Typic Argiborolls except
for *a* and part of or all of *f(1)*, with or without *b(1)* or *h*,
or both.
Abruptic Udic Argiborolls are like Typic Argiborolls except
for *a* and *f(2)*, with or without *b(1)* or *h*, or both.
Albic Argiborolls are like Typic Argiborolls except for *a*,
b(1), and *d*; or for *a* and *d*, with or without *f(2)*.
Aquic Argiborolls are like Typic Argiborolls except for *d*,
with or without *f(2)*.
Aridic Argiborolls are like Typic Argiborolls except for *f(1)*.
Boralfic Argiborolls are like Typic Argiborolls except for
b(2).
Boralfic Udic Argiborolls are like Typic Argiborolls except
for *b(2)* and *f(2)*.
Lithic Argiborolls are like Typic Argiborolls except for *e* or
for *e* and *f(1)*.
Pachic Argiborolls are like Typic Argiborolls except for *g*
with or without *d* or *f(1)* or both.
Pachic Udic Argiborolls are like Typic Argiborolls except
for *f(2)* and *g*, with or without *d*.
Udic Argiborolls are like Typic Argiborolls except for *f(2)*.
Ustertic Argiborolls are like Typic Argiborolls except for *h*
and *f(1)*, with or without *d* or *g*, or both.
Vertic Argiborolls are like Typic Argiborolls except for *h*,
with or without *d* or *g*, or both.

Calciborolls

Distinctions between Typic Calciborolls and other subgroups

Typic Calciborolls are the Calciborolls that
a. Do not have mottles that have chroma of 2 or less within
1 m of the surface if artificially drained or if the mottled
horizon is continuously saturated with water for as long as
90 days;
b. Have either or both
(1) A color value, dry, of less than 4.5 in the upper 18
cm of the mollic epipedon after mixing, or in any Ap
horizon that is >18 cm thick; or
(2) A moisture control section that is dry in some part
less than six-tenths of the time in most years that the soil
temperature at a depth of 50 cm is above 5°C;
c. Do not have a lithic contact within 50 cm of the surface;
and
d. Do not have a petrocalcic horizon that has its upper
boundary within 1 m of the surface.
Aridic Calciborolls are like Typic Calciborolls except for *b*.
Lithic Calciborolls are like Typic Calciborolls except for *c*
or for *b*(1) and *c*.
Petrocalcic Calciborolls are like Typic Calciborolls except
for *d*.

Cryoborolls

Distinctions between Typic Cryoborolls and other subgroups

Typic Cryoborolls are the Cryoborolls that
a. Do not have an argillic horizon;
b. Do not have a layer in the upper 75 cm that has texture finer than loamy fine sand, that is as much as 18 cm thick, that has a bulk density (at 33 kPa water tension) of 0.95 g per cubic centimeter or less in the fine-earth fraction, and that has either of the following:
    (1) A ratio of measured clay to 1500 kPa water (percentages) of 1.25 or less; or
    (2) A ratio of CEC (at pH near 8) to 1500 kPa water of >1.5 and more exchange acidity than the sum of bases plus KCl-extractable aluminum;
c. Do not have distinct or prominent mottles that are due to segregation of iron or manganese within 1 m of the surface if artificially drained, or if undrained, are not continuously saturated with water within a depth of 1 m for as long as 90 days;
d. Do not have a calcic horizon within or immediately under the mollic epipedon unless the lower part of the mollic epipedon is also an argillic horizon;
e. Have a mollic epipedon that is <40 cm thick or that has texture of loamy fine sand or coarser;
f. Have a mollic epipedon that is continuous throughout each pedon;
g. Do not have a lithic contact within 50 cm of the surface;
h. Have a mean annual soil temperature higher than 0°C;
i. Do not have an albic horizon immediately below the mollic epipedon;
j. Do not have a duripan that has its upper boundary within 1 m of the soil surface;
k. Have a regular decrease in organic carbon content with increasing depth and unless a lithic or a paralithic contact is at some depth between 50 cm and 1.25 m below the soil surface, have an organic carbon content of 0.3 percent or less at a depth within 1.25 m of the surface; or the slope is >25 percent; and
l. Do not have the following combination of characteristics:
    (1) Cracks at some period in most years that are 1 cm or more wide at a depth of 50 cm, that are at least 30 cm long in some part, and that extend upward to the surface or to the base of an Ap horizon; and
    (2) A potential linear extensibility of 6 cm or more in the upper 1 m of the soil or the whole soil if a lithic or paralithic contact is deeper than 50 cm but shallower than 1 m; and
    (3) More than 35 percent clay in horizons that total more than 50 cm in thickness.

Abruptic Cryoborolls are like Typic Cryoborolls except for *a* and *i* and have an increase in clay content of 20 percent (absolute) or more within a vertical distance of 7.5 cm below the upper boundary of the argillic horizon.
Andeptic Cryoborolls are like Typic Cryoborolls except for *b* or for *b* and *e*.
Aquic Cryoborolls are like Typic Cryoborolls except for *c*.
Argiaquic Cryoborolls are like Typic Cryoborolls except for *g* and *c*.

Argic Cryoborolls are like Typic Cryoborolls except for *a*, and the argillic horizon is continuous throughout each pedon.

Argic Lithic Cryoborolls are like Typic Cryoborolls except for *a* and *g*, with or without *e*, and the argillic horizon is continuous throughout each pedon.

Argic Pachic Cryoborolls are like Typic Cryoborolls except for *a* and *e*, with or without *c*.

Argic Vertic Cryoborolls are like Typic Cryoborolls except for *a* and *l*, with or without any or all of *c*, *e*, and *k*.

Boralfic Cryoborolls are like Typic Cryoborolls except for *a* and *i*, with or without *e*.

Boralfic Lithic Cryoborolls are like Typic Cryoborolls except for *a*, *g*, and *i*, with or without *d* or *e*, or both.

Calcic Cryoborolls are like Typic Cryoborolls except for *d*.

Calcic Pachic Cryoborolls are like Typic Cryoborolls except for *d* and *e*, with or without *a* or *c*, or both.

Cumulic Cryoborolls are like Typic Cryoborolls except for *e* and *k* or for *c*, *e*, and *k*.

Duric Cryoborolls are like Typic Cryoborolls except for *j* or for *j* and *a*.

Lithic Cryoborolls are like Typic Cryoborolls except for *g*, with or without any or all of *d*, *e*, and *h*.

Lithic Ruptic-Argic Cryoborolls are like Typic Cryoborolls except for *a* and *g*, and the argillic horizon is intermittent in each pedon.

Lithic Ruptic-Entic Cryoborolls are like Typic Cryoborolls except for *g* and *f*, with or without *h*.

Natric Cryoborolls are like Typic Cryoborolls except for *a*, with or without *i*, and have an SAR $\geq$13 (or $\geq$15 percent saturation with exchangeable sodium) in the major part of the argillic horizon.

Pachic Cryoborolls are like Typic Cryoborolls except for *e*, with or without *c*.

Pergelic Cryoborolls are like Typic Cryoborolls except for *h*.

Vertic Cryoborolls are like Typic Cryoborolls except for *l*, with or without all or any of *c*, *e*, or *k*.

Haploborolls

Distinctions between Typic Haploborolls and other subgroups

a. Do not have a layer in the upper 75 cm that has texture finer than loamy fine sand, that is as much as 18 cm thick, that has a bulk density (at 33 kPa water tension) of 0.95 g per cubic centimeter or less in the fine-earth fraction, and that has either of the following:

    (1) A ratio of measured clay to 1500 kPa water (percentages) of 1.25 or less; or

    (2) A ratio of CEC (at pH near 8) to 1500 kPa water of >1.5 and more exchange acidity than the sum of the bases plus KCl-extractable aluminum;

b. Do not have mottles that have chroma of 2 or less within 1 m of the surface if artificially drained, or if undrained, are not continuously saturated with water in the mottled horizon for as long as 90 days in most years;

c. Have both the following:

    (1) Either or both:

        (a) A color value, dry, of less than 4.5 in the upper 18 cm of the mollic epipedon, after mixing, or in any Ap horizon that is >18 cm thick; or

(b) A moisture control section that is dry in some part less than six-tenths of the time that the soil temperature at a depth of 50 cm is above 5°C in most years; and

(2) A chroma, moist, after rubbing of 1.5 or more in the upper part of the mollic epipedon after it has been mixed to a depth of 18 cm or in any Ap horizon that is >18 cm thick, and the soil is dry in all parts of the moisture control section at some time in most years;

d. Have a cambic horizon, or the lower part of the mollic epipedon meets the requirements of a cambic horizon except for color and organic-carbon content;

e. Have a regular decrease in organic carbon content with increasing depth and unless a lithic or a paralithic contact is at some depth between 50 cm and 1.25 m below the soil surface, have an organic carbon content of 0.3 percent or less at a depth within 1.25 m of the surface; or the slope is >25 percent; and

f. Do not have a lithic contact within 50 cm of the surface;

g. Do not have the following combination of characteristics:

(1) Cracks at some period in most years that are 1 cm or more wide at a depth of 50 cm and that are at least 30 cm long in some part and that extend upward to the surface or to the base of an Ap horizon;

(2) Potential linear extensibility of 6 cm or more in the upper 1 m of the soil or in the whole soil if a lithic or paralithic contact is deeper than 50 cm but shallower than 1 m; and

(3) More than 35 percent clay in horizons that total >50 cm in thickness;

h. Do not have a salic horizon that has its upper boundary within 75 cm of the surface; and

i. Have a mollic epipedon <40 cm thick, or the epipedon has a sandy particle-size class in the major part, or there is a paralithic contact or a sandy contrasting layer between depths of 40 and 50 cm.

Aquic Haploborolls are like Typic Haploborolls except for *b*, with or without *c* or *d*, or both.

Aridic Haploborolls are like Typic Haploborolls except for all or part of *c(1)*.

Cumulic Haploborolls are like Typic Haploborolls except for *i* and *e*, with or without all or any of *b*, *d*, and *c(1)*, and have a concave shape.

Cumulic Udic Haploborolls are like Typic Haploborolls except for *i*, *e*, and *c(2)*, with or without *b* or *d*, or both, and have a concave shape.

Entic Haploborolls are like Typic Haploborolls except for *d*.

Fluvaquentic Haploborolls are like Typic Haploborolls except for *b* and *e*, with or without either *d* or *c*, or both.

Fluventic Haploborolls are like Typic Haploborolls except for *e*, with or without *d*.

Lithic Haploborolls are like Typic Haploborolls except for *f*, with or without any or all of *c*, *d*, or *i*.

Pachic Haploborolls are like Typic Haploborolls except for *i*, with or without any or all of *b*, *d*, or *c(1)*.

Pachic Udic Haploborolls are like Typic Haploborolls except for *i* and *c(2)*, with or without *b* or *d*, or both.

Ruptic-Lithic Haploborolls are like Typic Haploborolls except for *f*, with or without *d*, and bedrock is within a depth of 50 cm in part of each pedon.

Torrifluventic Haploborolls are like Typic Haploborolls except for *e* and *c(1)*, with or without *d*.
Torriorthentic Haploborolls are like Typic Haploborolls except for all or part of *c(1)* and *d*.
Udertic Haploborolls are like Typic Haploborolls except for *c(2)* and *g*, with or without any or all or *c(1)(b)*, *e*, or *i*.
Udic Haploborolls are like Typic Haploborolls except for *c(2)*.
Udorthentic Haploborolls are like Typic Haploborolls except for *c(2)* and *d*.
Vertic Haploborolls are like Typic Haploborolls except for *g*, with or without any or all of *c(1)(b)*, *e*, or *i*.

Natriborolls

Distinctions between Typic Natriborolls and other subgroups

Typic Natriborolls are the Natriborolls that
a. Do not have tonguing or interfingering of an albic horizon more than 2.5 cm into the natric horizon;
b. Have
    (1) Either or both
        (a) A color value, dry, of less than 4.5 in the upper 18 cm of the mollic epipedon, after mixing, or in any Ap horizon that is >18 cm; or
        (b) A moisture control section that is dry in some part less than six-tenths of the time that the soil temperature at a depth of 50 cm is above 5°C in most years; or
    (2) A chroma, moist, after rubbing, of 1.5 or more in the upper part of the mollic epipedon to a depth of 18 cm, after mixing, or in any Ap horizon that is more than 18 cm thick; and
c. Do not have visible crystals or nests of gypsum or more soluble salts within 40 cm of the surface of the soil.
Aridic Natriborolls are like Typic Natriborolls except for *b(1)*.
Glossic Natriborolls are like Typic Natriborolls except for *a*.
Glossic Udic Natriborolls are like Typic Natriborolls except for *a* and *b(2)*.
Leptic Natriborolls are like Typic Natriborolls except for *c* or for *b* and *c*.
Udic Natriborolls are like Typic Natriborolls except for *b(2)*.

Paleborolls

Distinctions between Typic Paleborolls and other subgroups

Typic Paleborolls are the Paleborolls that
a. Do not have an argillic horizon that has an increase in clay content of 20 percent (absolute) or more within a vertical distance of 7.5 cm below its upper boundary;
b. Do not have mottles that have chroma of 2 or less within 1 m of the surface if artificially drained or, if undrained, are not continuously saturated with water in the mottled horizon for as long as 90 days in most years;
c. Have a mean summer soil temperature at a depth of 50 cm or at a lithic or paralithic contact, whichever is shallower, of 15°C or higher if there is no O horizon and 8°C or higher if there is an O horizon; and
d. Have a mollic epipedon that is <50 cm thick.

Abruptic Cryic Paleborolls are like Typic Paleborolls except for *a* and *c*, with or without *b*.
Cryic Paleborolls are like Typic Paleborolls except for *c*.
Cryic Pachic Paleborolls are like Typic Paleborolls except for *c* and *d*, with or without *b*.
Pachic Paleborolls are like Typic Paleborolls except for *d*, with or without *b*.

Vermiborolls

Distinctions between Typic Vermiborolls and other subgroups

Typic Vermiborolls are the Vermiborolls that
a. Have a mollic epipedon 75 cm or more thick;
b. Have
    (1) Either or both
        (a) A color value, dry, of less than 4.5 in the upper 18 cm of the mollic epipedon, after mixing, or in any Ap horizon that is >18 cm thick; or
        (b) A moisture control section that is dry in some part less than six-tenths of the time that the soil temperature at a depth of 50 cm is above 5°C in most years; and
    (2) A chroma, moist, after rubbing, of 1.5 or more in the upper part of the mollic epipedon to a depth of 18 cm, after mixing, or in any Ap horizon that is more than 18 cm thick; and
c. Do not have a lithic contact within 50 cm of the surface.
Haplic Vermiborolls are like Typic Vermiborolls except for *a*.
Hapludic Vermiborolls are like Typic Vermiborolls except for *a* and *b(2)*.

Rendolls

Distinctions between Typic Rendolls and other subgroups

Typic Rendolls are the Rendolls that
a. Have a soil temperature regime warmer than cryic;
b. Do not have a cambic horizon throughout the pedon;
c. Do not have a lithic contact within 50 cm of the surface; and
d. Do not have the following combination of characteristics:
    (1) Cracks at some period in most years that are 1 cm or more wide at a depth of 50 cm, that are at least 30 cm long in some part, and that extend upward to the surface or to the base of an Ap horizon, and
    (2) A coefficient of linear extensibility (COLE) of 0.09 or more in a horizon or horizons at least 50 cm thick and a potential linear extensibility of 6 cm or more in the upper 1 m of the soil or in the whole soil if a lithic or paralithic contact is deeper than 50 cm but shallower than 1 m, and
    (3) More than 35 percent clay in horizons that total >50 cm in thickness.
e. Have a dry color value of 5.5 or less after the surface soil to a depth of 18 cm has been mixed or of any Ap horizon that is deeper than 18 cm.
Cryic Rendolls are like Typic Rendolls except for *a*.

Cryic Lithic Rendolls are like Typic Rendolls except for *a* and *c* or for *a*, *b*, and *c*.

Entic Rendolls are like Typic Rendolls except for *e*.

Eutrochreptic Rendolls are like Typic Rendolls except for *b*, and the mean summer and mean winter soil temperatures at a depth of 50 cm differ by 5°C or more.

Eutropeptic Rendolls are like Typic Rendolls except for *b* and the mean summer and mean winter soil temperatures at a depth of 50 cm differ by <5°C.

Lithic Rendolls are like Typic Rendolls except for *c*.

Vertic Rendolls are like Typic Rendolls except for *d*.

Udolls

Key to great groups

GGA. Udolls that have an argillic horizon and a clay distribution such that the clay content does not decrease by as much as 20 percent of the maximum clay content within 1.5 m of the soil surface and there is no lithic or paralithic contact within that depth, and there is one or both of the following features:

1. Hue redder than 10YR and chroma greater than 4 dominant in the matrix in at least the lower part of an argillic horizon; or

2. Many coarse mottles that have hue redder than 7.5YR or chroma greater than 5.

Paleudolls, p. 186

GGB. Other Udolls that have an argillic horizon.

Argiudolls, p. 184

GGC. Other Udolls that have a mollic epipedon that, below any Ap horizon, is 50 percent or more by volume wormholes, wormcasts, or filled animal burrows and that either rests on a lithic contact or has a transition to an underlying horizon in which 25 percent or more of the material is discrete wormholes, wormcasts, or filled animal burrows that contain material from the mollic epipedon and from the underlying horizon.

Vermudolls, p. 186

GGD. Other Udolls.

Hapludolls, p. 185

Argiudolls

Distinctions between Typic Argiudolls and other subgroups

Typic Argiudolls are the Argiudolls that
a. Do not have mottles within 50 cm of the surface if the mottled horizon is saturated with water at some period of the year when the soil temperature in the mottled horizon is above 5°C or if the soil is artificially drained; and have a horizon 15 cm or more thick immediately below the mollic epipedon that either

    (1) Has a hue of 10YR or redder and chroma of 3 or more, and does not have mottles that have chroma of 2 or less and value of 4 or more if the mottled horizon is saturated with water at some period of the year when the

soil temperature in the mottled horizon is above 5°C or if the soil is artificially drained; or

(2) Has a hue of 2.5Y or redder and chroma of 4 or more;

b. Do not have a lithic contact within 50 cm of the surface;

c. Have texture finer than loamy fine sand in the argillic horizon, or the argillic horizon does not consist entirely of lamellae with a combined thickness of <15 cm;

d. Do not have the following combination of characteristics:

(1) Cracks at some period in most years that are 1 cm or more wide at a depth of 50 cm, that are at least 30 cm long in some part, and that extend upward to the surface or to the base of an Ap horizon,

(2) A coefficient of linear extensibility (COLE) of 0.09 or more in a horizon or horizons at least 50 cm thick and a potential linear extensibility of 6 cm or more in the upper 1 m of the soil or in the whole soil if a lithic or paralithic contact is deeper than 50 cm but shallower than 1 m, and

(3) More than 35 percent clay in horizons that total >50 cm in thickness; and

e. Have CEC (by 1 M NH$_4$OAc pH7) of 24 or more in the major part of the argillic horizon or the major part of the upper 100 cm of the argillic horizon if the argillic horizon is thicker than 100 cm.

Aquic Argiudolls are like Typic Argiudolls except for *a.*
Lithic Argiudolls are like Typic Argiudolls except for *b.*
Oxic Argiudolls are like Typic Argiudolls except for *e.*
Psammentic Argiudolls are like Typic Argiudolls except for *c.*
Vertic Argiudolls are like Typic Argiudolls except for *d* with or without *a.*

Hapludolls

Distinctions between Typic Hapludolls and other subgroups

Typic Hapludolls are the Hapludolls that

a. Do not have mottles within 40 cm of the surface if the mottled horizon is saturated with water at some period of the year when the soil temperature in the mottled horizon is above 5°C or if the soil is artificially drained; and have a horizon 15 cm or more thick immediately below the mollic epipedon that either

(1) Has a hue of 10YR or redder and chroma of 3 or more, and does not have mottles that have chroma of 2 or less and value of 4 or more if the mottled horizon is saturated with water at some period of the year when the soil temperature in the mottled horizon is above 5°C or if the soil is artificially drained, or

(2) Has a hue of 2.5 Y or redder and chroma of 4 or more;

b. Have a mollic epipedon <60 cm thick or texture that is loamy fine sand or coarser if the mollic epipedon is ≥60 cm thick;

c. Have a cambic horizon, or the lower part of the mollic epipedon meets the requirements of a cambic horizon except for color value and organic-carbon content, and either the cambic horizon or the lower part of the epipedon is free of carbonates in some part;

d. Have a regular decrease in organic carbon content with increasing depth and unless a lithic or a paralithic contact is at some depth between 50 cm and 1.25 m below the soil surface, have an organic carbon content of 0.3 percent or less at a depth within 1.25 m of the surface; or the slope is >25 percent;

e. Do not have a lithic contact within 50 cm of the surface; and

f. Do not have the following combination of characteristics;
  (1) Cracks at some period in most years that are 1 cm or more wide at a depth of 50 cm, that are at least 30 cm long in some part, and that extend upward to the surface or to the base of an Ap horizon;
  (2) A coefficient of linear extensibility (COLE) of 0.09 or more in a horizon or horizons at least 50 cm thick and a potential linear extensibility of 6 cm or more in the upper 1 m of the soil or in the whole soil if a lithic or paralithic contact is deeper than 50 cm but shallower than 1 m; and
  (3) More than 35 percent clay in horizons that total >50 cm in thickness.

Aquic Hapludolls are like Typic Hapludolls except for *a*, or *a* and *c*.

Cumulic Hapludolls are like Typic Hapludolls except for *b* and *d*, with or without *a* or *c*, or both.

Entic Hapludolls are like Typic Hapludolls except for *c*.

Fluvaquentic Hapludolls are like Typic Hapludolls except for *a* and *d* or for *a*, *c*, and *d*.

Fluventic Hapludolls are like Typic Hapludolls except for *d* or for *c* and *d*.

Lithic Hapludolls are like Typic Hapludolls except for *e*, with or without *a* or *c*, or both.

Vermic Hapludolls are like Typic Hapludolls except for *b* and *c* and have mollic epipedon that, below any Ap horizon, has 50 percent or more by volume of wormholes, wormcasts, or filled animal burrows.

Vertic Hapludolls are like Typic Hapludolls except for *f*, with or without all or any of *a*, *b*, and *d*.

Paleudolls

Distinctions between Typic Paleudolls and other subgroups

Typic Paleudolls are the Paleudolls that
a. Do not have mottles that have chroma of 2 or less in the upper 50 cm of the argillic horizon if the mottled horizon is saturated with water at some period when its temperature is >5°C or if the soil has artificial drainage.

Aquic Paleudolls are like Typic Paleudolls except for *a*.

Vermudolls

Distinctions between Typic Vermudolls and other subgroups

Typic Vermudolls are the Vermudolls that
a. Have a mollic epipedon that is 75 cm or more thick;
b. Do not have a cambic horizon;
c. Have a mollic epipedon that has a transition to the underlying horizon in which 50 percent or more of the material is discrete wormholes, wormcasts, or animal

burrows filled with material from the mollic epipedon and
the underlying horizon;
d. Do not have a lithic contact within 50 cm of the surface;
and
e. Have a mollic epipedon that below any Ap horizon has
granular structure formed mainly from wormholes,
wormcasts, or filled animal burrows.
Entic Vermudolls are like Typic Vermudolls except for *a*.
Haplic Vermudolls are like Typic Vermudolls except for *b*
and *c*, with or without *a*.
Lithic Vermudolls are like Typic Vermudolls except for *d*
and *a*, with or without *c*.

Ustolls

Key to great groups

GFA. Ustolls that have a duripan with its upper boundary
within 1 m of the soil surface.

Durustolls, p. 190

GFB. Other Ustolls that have a natric horizon.

Natrustolls, p. 193

GFC. Other Ustolls that have a petrocalcic horizon that has
its upper boundary within 1.5 m of the soil surface, and that
have an argillic horizon or are noncalcareous in some
subhorizon above the petrocalcic horizon after the surface
soil to a depth of 18 cm has been mixed, or that have an
argillic horizon that has one or both of the following:
  1. A vertical clay distribution such that the clay content
  does not decrease by as much as 20 percent of the
  maximum clay content within 1.5 m of the soil surface
  and the soil does not have a lithic or paralithic contact
  within that depth, and the argillic horizon has one or
  both of these:
    a. A hue redder than 10YR and chroma higher than 4
    in the matrix; or
    b. Common coarse mottles that have a hue of 7.5YR
    or redder or chroma higher than 5; or
  2. A particle-size class in the upper part that is clayey
  and an increase of at least 20 percent clay (absolute)
  within a vertical distance of 7.5 cm or of 15 percent clay
  (absolute) within 2.5 cm at the upper boundary, and there
  is no lithic or paralithic contact within 50 cm of the
  surface of the soil.

Paleustolls, p. 193

GFD. Other Ustolls that do not have an argillic horizon
above a calcic, gypsic, or petrocalcic horizon, and that have
a calcic or gypsic horizon that has its upper boundary within
1.5 m of the surface, and that are calcareous in all overlying
subhorizons after the upper soil to a depth of 18 cm has
been mixed, unless the texture is coarser than loamy very
fine sand or very fine sand.

Calciustolls, p. 189

GFE. Other Ustolls that have an argillic horizon.

Argiustolls, p. 188

GFF. Other Ustolls that have a mollic epipedon below any Ap horizon that is 50 percent or more by volume wormholes and wormcasts or filled animal burrows, and that either rests on a lithic contact or has a transition to the underlying horizon in which 25 percent or more of the material is discrete wormcasts or animal burrows filled with material from the mollic epipedon and the underlying horizon.

Vermustolls, p. 194

GFG. Other Ustolls.

Haplustolls, p. 191

Argiustolls

Distinctions between Typic Argiustolls and other subgroups

Typic Argiustolls are the Argiustolls that
a. Do not have mottles that have chroma of 2 or less within 1 m of the soil surface if artificially drained or, if undrained, are not continuously saturated with water within 1 m of the soil surface for as long as 3 months in most years;
b. Do not have a brittle horizon 15 cm or more thick within 1 m of the soil surface that contains some opal coatings or some (<20 percent by volume) durinodes;
c. Do not have a lithic contact within 50 cm of the surface;
d. Do not have an albic horizon or other eluvial horizon above the argillic horizon that has a color value too high for a mollic epipedon and chroma too high for an albic horizon;
e. Have a mollic epipedon <50 cm thick, or the texture is loamy fine sand or coarser if the mollic epipedon is >50 cm thick;
f. When neither irrigated nor fallowed to store moisture:
  (1) If the soil temperature regime is mesic or thermic, are dry for more than four-tenths of the cumulative days in some part of the moisture control section when the soil temperature at a depth of 50 cm exceeds 5°C; or
  (2) If the soil temperature regime is hyperthermic, isomesic, or warmer, the soils are dry in some or all parts of the moisture control section for more than 90 days during a period when the soil temperature at a depth of 50 cm exceeds 8°C;
g. Do not have the following combination of characteristics:
  (1) Cracks at some period in most years that are 1 cm or more wide at a depth of 50 cm, that are at least 30 cm long in some part, and that extend upward to the surface or to the base of an Ap horizon,
  (2) A coefficient of linear extensibility (COLE) of 0.07 or more in a horizon or horizons at least 50 cm thick and a potential linear extensibility of 6 cm or more in the upper 1.25 m of the soil or in the whole soil if a lithic or paralithic contact is deeper than 50 cm but shallower than 1.25 m, and
  (3) More than 35 percent clay in horizons that total >50 cm in thickness;
h. When neither irrigated nor fallowed to store moisture,
  (1) If the soil temperature regime is mesic or thermic, are dry less than six-tenths of the time in half or more years in some part of the moisture control section (not necessarily the same part) during a period when the soil temperature at a depth of 50 cm is higher than 5°C, or

(2) If the soil temperature regime is hyperthermic, or isomesic, or warmer, are moist in some or all parts of the moisture control section for 90 consecutive days or more during a period when the soil temperature at a depth of 50 cm is higher than 8°C.

Alfic Lithic Argiustolls are like Typic Argiustolls except for *c* and *d*, with or without *f* or *h*, or both.

Aquic Argiustolls are like Typic Argiustolls except for *a* or for *a* and *f*.

Aridic Argiustolls are like Typic Argiustolls except for *h*.

Boralfic Argiustolls are like Typic Argiustolls except for *d*, with or without all or any of *a*, *f*, or *h*, and the mean annual soil temperature is lower than 10°C.

Lithic Argiustolls are like Typic Argiustolls except for *c*, with or without *f* or *h*, or both.

Pachic Argiustolls are like Typic Argiustolls except for *e*, with or without *a* or *f*, or both.

Torrertic Argiustolls are like Typic Argiustolls except for *g*, with or without *e* or *h*, and the cracks are open 6 months or more in most years.

Udertic Argiustolls are like the Typic Argiustolls except for *f* and *g*, with or without *a* or *e*, or both, and cracks are open less than 135 days in most years.

Udic Argiustolls are like Typic Argiustolls except for *f*.

Ustalfic Argiustolls are like Typic Argiustolls except for *d* with or without *a*, *f*, and *h*, and the mean soil temperature is 10°C or more.

Vertic Argiustolls are like Typic Argiustolls except for *g* with or without *a* or *e* or both, and cracks are open between 135 and 180 days in most years.

Calciustolls

Distinctions between Typic Calciustolls and other subgroups

Typic Calciustolls are the Calciustolls that
a. Do not have mottles within 75 cm of the surface that are due to segregation of iron or manganese accompanied by seasonal ground water and, if undrained, are not continuously saturated with water for as long as 90 days within 1 m of the surface;
b. When neither irrigated nor fallowed to store moisture
    (1) If the soil temperature regime is mesic or thermic, are dry less than six-tenths of the time in half or more years in some part of the moisture control section (not necessarily the same part) during a period when the soil temperature at a depth of 50 cm exceeds 5°C, or
    (2) If the soil temperature regime is hyperthermic or isomesic, or warmer, are moist in some or all parts of the moisture control section for 90 consecutive days or more during a period when the soil temperature at a depth of 50 cm exceeds 8°C;
c. Do not have a lithic contact within 50 cm of the soil surface;
d. Have a mollic epipedon that is <50 cm thick, or the texture is loamy fine sand or coarser if the mollic epipedon is ≥50 cm thick;
e. Do not have a petrocalcic horizon that has its upper boundary within 1 m of the surface;
f. Do not have a salic horizon that has its upper boundary within 75 cm of the surface; and

g. Do not have the following combination of characteristics:
(1) Cracks at some period in most years that are 1 cm or more wide at a depth of 50 cm, that are at least 30 cm long in some part, and that extend upward to the surface or to the base of an Ap horizon,
(2) A coefficient of linear extensibility (COLE) of 0.07 or more in a horizon or horizons at least 50 cm thick and a potential linear extensibility of 6 cm or more in the upper 1.25 m of the soil or in the whole soil if a lithic or paralithic contact is deeper than 50 cm but shallower than 1.25 m, and
(3) More than 35 percent clay in horizons that total >50 cm in thickness.
h. When neither irrigated nor fallowed to store moisture:
(1) If the soil temperature regime is mesic or thermic, are dry for more than four-tenths of the cumulative days in some part of the moisture control section when the soil temperature at a depth of 50 cm exceeds 5°C; or
(2) If the soil temperature regime is hyperthermic, isomesic, or warmer, the soils are dry in some or all parts of the moisture control section for more than 90 days during a period when the soil temperature at a depth of 50 cm exceeds 8°C;

Aquic Calciustolls are like Typic Calciustolls except for *a* or for *a* and *b*.
Aridic Calciustolls are like Typic Calciustolls except for *b*.
Lithic Calciustolls are like Typic Calciustolls except for *c* or for *b* and *c*.
Lithic Petrocalcic Calciustolls are like Typic Calciustolls except for *c* and *e* or for *c*, *e*, or *b*.
Pachic Calciustolls are like Typic Calciustolls except for *d*, with or without *a*.
Petrocalcic Calciustolls are like Typic Calciustolls except for *e* or for *b* and *e*.
Torrertic Calciustolls are like Typic Calciustolls except for *g*, with or without any or all of *a*, *b*, and *d*, and the cracks are open 180 days or more, cumulative, in most years.
Udertic Calciustolls are like Typic Calciustolls except for *g* and *h*, with or without *a* or *b* or both, and cracks are open less than 135 days in most years.
Udic Calciustolls are like Typic Calciustolls except for *h*.
Vertic Calciustolls are like Typic Calciustolls except for *g*, with or without any or all of *a*, *b*, and *d*, and the cracks are open between 135 and 180 days in most years.

Durustolls

Distinctions between Typic Durustolls and other subgroups.

Typic Durustolls are the Durustolls that:
a. Have an argillic horizon above the duripan;
b. Have a duripan that is massive and platy and that has half or more of its upper boundary coated or indurated with opal and silica with or without sesquioxides or that is indurated in some subhorizon below its upper boundary;
c. Do not have a natric horizon above the duripan;
d. When neither irrigated nor fallowed to store moisture:
(1) If the soil temperature regime is mesic or thermic, are dry less then six-tenths of the time in half or more years in some part of the moisture control section (not

necessarily the same part) during a period when the soil temperature at a depth of 50 cm exceeds 5°C, or
(2) If the soil temperature regime is hyperthermic or isomesic, or warmer, are moist in some or all parts of the moisture control section for 90 consecutive days or more during a period when the soil temperature at a depth of 50 cm exceeds 8°C.

Aridic Durustolls are like Typic Durustolls except for *d* and have an aridic moisture regime that border on ustic.
Orthidic Durustolls are like Typic Durustolls except for *a* and *d* and have an aridic moisture regime that borders on ustic.

Haplustolls

Distinctions between Typic Haplustolls and other subgroups

Typic Haplustolls are the Haplustolls that
a. Do not have mottles that have chroma of 2 or less within 1 m of the surface if artificially drained or, if undrained, are not continuously saturated with water within 1 m of the soil surface for 90 days or more in most years;
b. Have a mollic epipedon that is <50 cm thick, or the texture is loamy fine sand or coarser if the mollic epipedon is ≥50 cm thick;
c. Do not have a brittle horizon 15 cm or more thick within 1 m of the surface that contains some opal coatings or some durinodes (<20 percent by volume);
d. Have a cambic horizon, or the lower part of the mollic epipedon meets the requirements of a cambic horizon except for color and for organic-carbon content, and either the cambic horizon or the lower part of the mollic epipedon is free of carbonates in some part;
e. Have a regular decrease in organic-carbon content with increasing depth to a level of 0.3 percent or less within 1.25 m of the surface unless a lithic or paralithic contact occurs at a shallower depth or the slope is >25 percent;
f. Do not have a lithic contact within 50 cm of the surface;
g. Do not have a salic horizon that has its upper boundary within 75 cm of the surface;
h. Do not have the following combination of characteristics:
(1) Cracks at some period in most years that are 1 cm or more wide at a depth of 50 cm, that are at least 30 cm long in some part, and that extend upward to the surface or to the base of an Ap horizon;
(2) A coefficient of linear extensibility (COLE) of 0.07 or more in a horizon or horizons at least 50 cm thick and a potential linear extensibility of 6 cm or more in the upper 1.25 m of the soil or in the whole soil if a lithic or paralithic contact is deeper than 50 cm but shallower than 1.25 m; or
(3) More than 35 percent clay in horizons that total >50 cm in thickness;
i. When neither irrigated nor fallowed to store moisture:
(1) If the soil temperature regime is mesic or thermic, are dry for more than four-tenths of the cumulative days in some part of the moisture control section when the soil temperature at a depth of 50 cm exceeds 5°C; or
(2) If the soil temperature regime is hyperthermic, isomesic, or warmer, the soils are dry in some or all parts of the moisture control section for more than 90 days

during a period when the soil temperature at a depth of 50 cm exceeds 8°C;

j. When neither irrigated nor fallowed to store moisture,

   (1) If the soil temperature regime is mesic or thermic, are dry less than six-tenths of the time in half or more years in some part of the moisture control section (not necessarily the same part) during a period when the soil temperature at a depth of 50 cm exceeds 5°C, or

   (2) If the soil temperature regime is hyperthermic or isomesic, or warmer, are moist in some or all parts of the moisture control section for 90 consecutive days or more during a period when the soil temperature at a depth of 50 cm exceeds 8°C; and

k. Have CEC (by 1 M NH$_4$OAc pH7) of 24 or more cmol(+) kg^{-1} clay in the major part of the soil below a depth of 25 cm but above 100 cm or a lithic or paralithic contact if one is shallower than 100 cm.

Aquic Haplustolls are like Typic Haplustolls except for *a*, with or without *d* or *i*, or both.

Aridic Haplustolls are like Typic Haplustolls except for *j*.

Cumulic Haplustolls are like Typic Haplustolls except for *b* and *e*, with or without all or any of *a*, *d*, *i*, and *j*.

Entic Haplustolls are like Typic Haplustolls except for *d* and either do not have a cambic horizon or the epipedon is calcareous.

Fluvaquentic Haplustolls are like Typic Haplustolls except for *a* and *e*, with or without *d* or *i*, or both.

Fluventic Haplustolls are like Typic Haplustolls except for *e*, with or without *d* or *i*, or both.

Lithic Haplustolls are like Typic Haplustolls except for *f*, with or without any or all of *d*, *i*, and *j*.

Lithic Ruptic-Entic Haplustolls are like Typic Haplustolls except for *f* and *d* and have a cambic horizon in some part but in less than half of each pedon.

Oxic Haplustolls are like Typic Haplustolls except for *k*, with or without *b* or *e*, or both.

Pachic Haplustolls are like Typic Haplustolls except for *b*, with or without all or any of *a*, *d*, or *i*.

Ruptic-Lithic Haplustolls are like Typic Haplustolls except for *f* in part of each pedon.

Salorthidic Haplustolls are like Typic Haplustolls except for *g*, with or without all or any of *a*, *b*, *e*, *i*, or *j*.

Torrertic Haplustolls are like Typic Haplustolls except for *j(1)*, with or without all or any of *b*, *d*, *e*, *i*, or *j*, and the cracks are open more than 6 months in most years.

Torrifluventic Haplustolls are like Typic Haplustolls except for *e* and *j*, with or without *d*.

Torriorthentic Haplustolls are like Typic Haplustolls except for *d* and *j*.

Torroxic Haplustolls are like Typic Haplustolls except for *j* and *k*, with or without *b* or *e*.

Udertic Haplustolls are like Typic Haplustolls except for *h(1)* and *i*, with or without all or any of *a*, *b*, *d*, *e*, or *j*, and the cracks are open less than 135 days in most years.

Udic Haplustolls are like Typic Haplustolls except for *i*.

Udorthentic Haplustolls are like Typic Haplustolls except for *d* and *i*.

Vertic Haplustolls are like Typic Haplustolls except for *h(1)*, with or without all or any of *a*, *b*, *d*, *e*, or *j*, and the cracks are open between 135 and 180 days in most years.

Natrustolls

Distinctions between Typic Natrustolls and other subgroups

Typic Natrustolls are the Natrustolls that
a. Do not have any of the following characteristics within 1 m of the surface:
(1) Dominant chroma of 1 or less throughout and hue as yellow or yellower than 2.5 Y in some part;
(2) Dominant chroma of 2 or less and mottles that are not due to segregated lime; or
(3) Dominant chroma of 2 or less and a decrease in the percentage of exchangeable sodium from the upper 25-centimeter layer to the underlying layer;
b. Do not have a brittle horizon 15 cm or more thick that is within 1 m of the surface and that contains some opal coatings or some durinodes (<20 percent by volume);
c. Do not have tonguing or interfingering of an albic horizon more than 2.5 cm into a natric horizon;
d. When neither irrigated nor fallowed to store moisture
(1) If the soil temperature regime is mesic or thermic, are dry less than six-tenths of the time in half or more years in some part of the moisture control section (not necessarily the same part) during a period when the soil temperature at a depth of 50 cm exceeds 5°C, or
(2) If the soil temperature regime is hyperthermic, or isomesic or warmer, are moist in some or all parts of the moisture control section for 90 consecutive days or more during a period when the soil temperature at a depth of 50 cm exceeds 8°C; and
e. Do not have visible crystals or nests of gypsum or more soluble salts within 40 cm of the surface.
Aridic Natrustolls are like Typic Natrustolls except for *d.*
Glossic Natrustolls are like Typic Natrustolls except for *c.*
Leptic Natrustolls are like Typic Natrustolls except for *e* or for *d* and *e.*

Paleustolls

Distinctions between Typic Paleustolls and other subgroups

Typic Paleustolls are the Paleustolls that
a. Are noncalcareous in some horizons after the upper soil to a depth of 18 cm has been mixed;
b. Do not have mottles that have chroma of 2 or less within 1 m of the surface if artificially drained or, if undrained in most years, are not continuously saturated with water in the mottled horizon for as long as 90 days;
c. Have a mollic epipedon that is <50 cm thick or have texture that is loamy fine sand or coarser if the mollic epipedon is ≥50 cm thick;
d. Do not have a petrocalcic horizon within 1.5 m of the surface;
e. When neither irrigated nor fallowed to store moisture:
(1) If the soil temperature regime is mesic or thermic, are dry for more than four-tenths of the cumulative days in some part of the moisture control section when the soil temperature at a depth of 50 cm exceeds 5°C; or
(2) If the soil temperature regime is hyperthermic, isomesic, or warmer, the soils are dry in some or all parts of the moisture control section for more than 90 days

during a period when the soil temperature at a depth of 50 cm exceeds 8°C;

f. Do not have the following combination of characteristics:
(1) Cracks at some period in most years that are 1 cm or more wide at a depth of 50 cm, that are at least 30 cm long in some part, and that extend upward to the surface or to the base of an Ap horizon,
(2) A coefficient of linear extensibility (COLE) of 0.07 or more in a horizon or horizons at least 50 cm thick and a potential linear extensibility of 6 cm or more in the upper 1.25 m of the soil or in the whole soil if a lithic or paralithic contact is deeper than 50 cm but shallower than 1.25 m, and
(3) More than 35 percent clay in horizons that have total thickness of >50 cm; and

g. When neither irrigated nor fallowed to store moisture,
(1) If the soil temperature regime is mesic or thermic, are dry less than six-tenths of the time in half or more years in some part of the moisture control section (not necessarily the same part) during a period when the soil temperature at a depth of 50 cm exceeds 5°C, or
(2) If the soil temperature regime is hyperthermic or isomesic, or warmer, are moist in some or all parts of the moisture control section for 90 consecutive days or more during a period when the soil temperature at a depth of 50 cm exceeds 8°C.

Aquic Paleustolls are like Typic Paleustolls except for *b*.

Aridic Paleustolls are like Typic Paleustolls except for *g*.

Calcic Paleustolls are like Typic Paleustolls except for *a* and have a calcic horizon within a depth a 1 m if the particle-size class of the upper 50 cm of the argillic horizon is sandy, 60 cm if it is loamy, and 50 cm if it is clayey.

Calciorthidic Paleustolls are like Typic Paleustolls except for *a* and *g* and have a calcic horizon within a depth of 1 m if the particle-size class of the upper 50 cm of the argillic horizon is sandy, 60 cm if loamy, and 50 cm if clayey.

Pachic Paleustolls are like Typic Paleustolls except for *c*, with or without *b* or *e*, or both.

Petrocalcic Paleustolls are like Typic Paleustolls except for *d*, with or without *a* or *g*, or both, and have an argillic horizon.

Torrertic Paleustolls are like Typic Paleustolls except for *f*, with or without *c* or *g*, or both and the cracks are open more than 180 days, cumulative, in most years.

Udertic Paleustolls are like Typic Paleustolls except for *e* and *f*, with or without any or all of *b*, *c*, and *g*, and the cracks are open less than 135 days, cumulative, in most years.

Udic Paleustolls are like Typic Paleustolls except for *e*.

Vertic Paleustolls are like Typic Paleustolls except for *f*, with or without any or all of *b*, *c*, and *g*, and the cracks are open between 135 and 180 days, cumulative, in most years.

Vermustolls

Distinctions between Typic Vermustolls and other subgroups

Typic Vermustolls are the Vermustolls that
a. Have a mollic epipedon that is 50 cm or more thick but is <75 cm thick;
b. Do not have a cambic horizon;

c. Have a mollic epipedon that, below any Ap horizon, has a transition to the underlying horizon in which 50 percent or more of the volume is wormholes and wormcasts or filled animal burrows;
d. Do not have a lithic contact within 50 cm of the surface;
e. Have a mollic epipedon that has granular structure and is composed almost entirely, below any Ap horizon, of wormholes, wormcasts, or filled animal burrows; and
f. Do not have mottles that have chroma of 2 or less within 1 m of the surface.
Entic Vermustolls are like Typic Vermustolls except for *a*, and the epipedon is <50 cm thick.
Haplic Vermustolls are like Typic Vermustolls except for *b* and *c*, with or without *a*, and the epipedon is <75 cm thick.
Lithic Vermustolls are like Typic Vermustolls except for *d* and *a*, with or without *b* or *c*, or both, and the epipedon is <75 cm thick.
Pachic Vermustolls are like Typic Vermustolls except for *a*, and the epipedon is 75 cm or more thick.

Xerolls

Key to great groups

GDA. Xerolls that have a duripan within 1 m of the soil surface.

Durixerolls, p. 198

GDB. Other Xerolls that have a natric horizon but do not have a petrocalcic horizon that has its upper boundary within 1.5 m of the soil surface.

Natrixerolls, p. 200

GDC. Other Xerolls that have a petrocalcic horizon that has its upper boundary within 1.5 m of the soil surface or an argillic horizon that has either or both
    1. A vertical clay distribution such that the clay content does not decrease by as much as 20 percent of the maximum clay content within 1.5 m of the soil surface, and also one or more of
        a. A hue redder than 10YR and chroma higher than 4 in the matrix; or
        b. Common coarse mottles that have a hue of 7.5YR or redder or chroma higher than 5, or both; or
    2. A particle-size class in the upper part that is clayey and an increase in clay content of at least 20 percent clay (absolute) within a vertical distance of 7.5 cm or an increase of 15 percent clay (absolute) within a distance of 2.5 cm at the upper boundary and no lithic or paralithic contact within 50 cm of the soil surface.

Palexerolls, p. 200

GDD. Other Xerolls that have a calcic or gypsic horizon that has its upper boundary within 1.5 m of the soil surface and that are calcareous in all parts of all horizons above the calcic or gypsic horizon after the upper soil to a depth of 18 cm has been mixed unless the texture is coarser than loamy very fine sand or very fine sand.

Calcixerolls, p. 197

GDE. Other Xerolls that have an argillic horizon.

Argixerolls, p. 196

GDF. Other Xerolls.

Haploxerolls, p. 198

Argixerolls

Distinctions between Typic Argixerolls and other subgroups

Typic Argixerolls are the Argixerolls that
a. Do not have mottles that have chroma of 2 or less within 75 cm of the surface if artificially drained or if undrained are not continuously saturated with water within 1 m of the soil surface for 90 days or more in most years;
b. Do not have an albic horizon above the argillic horizon;
c. Do not have a calcic horizon or soft, powdery secondary lime within a depth of 1.5 m if the weighted average particle-size class of the upper 50 cm of the argillic horizon is sandy, 1.1 m if it is loamy, and 90 cm if it is clayey, or above a lithic contact that is shallower than these depths;
d. Do not have a horizon within 1 m of the surface that is >15 cm thick that either contains at least 20 percent durinodes in a nonbrittle matrix or is brittle and has firm consistence when moist;
e. Do not have a lithic contact within 50 cm of the soil surface;
f. Have a xeric moisture regime;
g. Have a mollic epipedon that is <50 cm thick or the texture is loamy fine sand or coarser if the mollic epipedon is ≥50 cm thick;
h. Have base saturation (by sum of cations) of >75 percent throughout the upper 75 cm or above a lithic or paralithic contact, whichever is shallower; and
i. Do not have the following combination of characteristics:
    (1) Cracks at some period in most years that are 1 cm or more wide at a depth of 50 cm, that are at least 30 cm long in some part, and that extend upward to the surface or to the base of an Ap horizon, and
    (2) A coefficient of linear extensibility (COLE) of 0.05 or more in a horizon or horizons at least 50 cm thick and a potential linear extensibility of 6 cm or more in the upper 1.5 m of the soil or in the whole soil if a lithic or paralithic contact is deeper than 50 cm but shallower than 1.5 m, and
    (3) More than 35 percent clay in horizons that have a total thickness of >50 cm.

Aquic Argixerolls are like Typic Argixerolls except for *a*, with or without *c*.

Aquultic Argixerolls are like Typic Argixerolls except for *a* and *h*.

Aridic Argixerolls are like Typic Argixerolls except for *f* and they have an aridic moisture regime that borders on xeric.

Aridic Calcic Argixerolls are like Typic Argixerolls except for *c* and *f*, and they have an aridic moisture regime that borders on xeric.

Boralfic Argixerolls are like Typic Argixerolls except for *b*, with or without all or any of *c*, *g*, or *h*, and the mean annual soil temperature is lower than 10°C.

Calcic Argixerolls are like Typic Argixerolls except for *c*.

Calcic Pachic Argixerolls are like Typic Argixerolls except for *c* and *g*, with or without *a*.
Durargidic Argixerolls are like Typic Argixerolls except for *d* and *f*.
Duric Argixerolls are like Typic Argixerolls except for *d*, with or without *c*.
Lithic Argixerolls are like Typic Argixerolls except for *e*, with or without *c* or *f*, or both.
Lithic Ultic Argixerolls are like Typic Argixerolls except for *e* and *h*.
Pachic Argixerolls are like Typic Argixerolls except for *g*, with or without *a*.
Pachic Ultic Argixerolls are like Typic Argixerolls except for *g* and *h*, with or without *a*.
Ultic Argixerolls are like Typic Argixerolls except for *h*.
Vertic Argixerolls are like Typic Argixerolls except for *i*, with or without all or any of *a*, *c*, or *g*.

Calcixerolls

Distinctions between Typic Calcixerolls and other subgroups

Typic Calcixerolls are the Calcixerolls that
a. Do not have mottles within 75 cm of the surface that are due to segregation of iron or manganese if artificially drained, and if undrained are not continuously saturated with water within 1 m of the soil surface for as long as 90 days in most years;
b. Have a mollic epipedon that is <50 cm thick, or the texture is loamy fine sand or coarser if the mollic epipedon is ≥ 50 cm thick;
c. Have a xeric moisture regime;
d. Do not have a lithic contact within 50 cm of the soil surface;
e. Do not have a mollic epipedon that below any Ap horizon has 50 percent or more by volume wormholes, wormcasts, or filled animal burrows; and
f. Do not have the following combination of characteristics:
   (1) Cracks at some period in most years that are 1 cm or more wide at a depth of 50 cm, that are at least 30 cm long in some part, and that extend upward to the surface or to the base of an Ap horizon;
   (2) A coefficient of linear extensibility (COLE) of 0.05 or more in a horizon or horizons at least 50 cm thick and a potential linear extensibility of 6 cm or more in the upper 1.5 m of the soil or in the whole soil if a lithic or paralithic contact is deeper than 50 cm but shallower than 1.5 m; and
   (3) More than 35 percent clay in horizons that have a total thickness of >50 cm.
Aquic Calcixerolls are like Typic Calcixerolls except for *a*.
Aridic Calcixerolls are like Typic Calcixerolls except for *c* and have an aridic moisture regime that borders on xeric.
Lithic Calcixerolls are like Typic Calcixerolls except for *d* or for *c* and *d*.
Pachic Calcixerolls are like Typic Calcixerolls except for *b* or for *a* and *b*, with or without *c*.
Vermic Calcixerolls are like Typic Calcixerolls except for *e*.
Vertic Calcixerolls are like Typic Calcixerolls except for *f*, with or without *a* or *b*, or both.

Durixerolls

Distinctions between Typic Durixerolls and other subgroups

Typic Durixerolls are the Durixerolls that
a. Do not have an argillic horizon that has an increase in
clay content of 20 percent (absolute) or more within a vertical distance of 7.5 cm or an increase of 15 percent or more
(absolute) within a distance of 2.5 cm at the upper boundary;
b. Have a duripan that is massive, platy, or prismatic and
that has half or more of its upper boundary indurated or
coated with opal or opal and sesquioxides or that is indurated in some subhorizon below its upper boundary;
c. Have an argillic horizon above the duripan;
d. Have a xeric moisture regime; and
e. Do not have mottles that have chroma of 2 or less above
the duripan.
Abruptic Aridic Durixerolls are like Typic Durixerolls except
for *a* and *d* and have an aridic moisture regime that borders
on xeric.
Argic Durixerolls are like Typic Durixerolls except for *b*.
Aridic Durixerolls are like Typic Durixerolls except for *d*
and have an aridic moisture regime that borders on xeric.
Entic Durixerolls are like Typic Durixerolls except for *b* and
c.
Haplic Durixerolls are like Typic Durixerolls except for *c*.
Orthidic Durixerolls are like Typic Durixerolls except for *c*
and *d* and have an aridic moisture regime that borders on
xeric.

Haploxerolls

Distinctions between Typic Haploxerolls and other subgroups

Typic Haploxerolls are the Haploxerolls that
a. Do not have mottles that have chroma of 2 or less within
75 cm of the surface if artificially drained and if undrained,
are not continuously saturated with water within 1 m of the
soil surface for 90 days or more in most years;
b. Do not have a calcic horizon or soft, powdery secondary
lime within a depth of 1.5 m if the weighted average particle-size class of all horizons between a depth of 25 cm and
1 m, or between a depth of 25 cm and a lithic or paralithic
contact that is shallower than 1 m, is sandy; within 1.1 m if
the average particle-size class is loamy; or within 90 cm if it
is clayey;
c. Have a mollic epipedon that is <50 cm thick or texture
that is loamy fine sand or coarser if the mollic epipedon is
≥50 cm thick;
d. Do not have a horizon within 1 m of the surface that is >
15 cm thick that either contains at least 20 percent durinodes in a nonbrittle matrix or is brittle and has firm consistence when moist;
e. Have a cambic horizon, or the lower part of the epipedon
meets the requirements of a cambic horizon except for color
value, and either the cambic horizon or the lower part of
the epipedon is free of carbonates in some part;
f. Have a regular decrease in organic carbon content with
increasing depth and unless a lithic or a paralithic contact is
at some depth between 50 cm and 1.25 m below the soil

surface, have an organic carbon content of 0.3 percent or less at a depth within 1.25 m of the surface; or the slope is >25 percent;

g. Do not have a lithic contact within 50 cm of the soil surface;

h. Have base saturation (by sum of cations) of >75 percent throughout the upper soil to a depth of 75 cm or above a lithic or paralithic contact, whichever is shallower;

i. Do not have a mollic epipedon that has granular structure and that, below any Ap horizon, has 50 percent or more by volume or wormholes, wormcasts, or filled animal burrows;

j. Do not have the following combination of characteristics:

(1) Cracks at some period in most years that are 1 cm or more wide at a depth of 50 cm, that are at least 30 cm long in some part, and that extend upward to the soil surface or to the base of an Ap horizon;

(2) A coefficient of linear extensibility (COLE) of 0.05 or more in a horizon or horizons at least 50 cm thick and a potential linear extensibility of 6 cm or more in the upper 1.5 m of the soil or in the whole soil if a lithic or paralithic contact is deeper than 50 cm but shallower than 1.5 m; and

(3) More than 35 percent clay in horizons that have a total thickness of >50 cm;

k. Have a xeric moisture regime.

Aquic Haploxerolls are like Typic Haploxerolls except for *a*, with or without *b* or *e*, or both.

Aquic Duric Haploxerolls a and *d*, with or without all or any of *b*, *e*, or *f*.

Aquultic Haploxerolls are like Typic Haploxerolls except for *a* and *h*, with or without *e*.

Aridic Haploxerolls are like Typic Haploxerolls except for *k* or for *h* and *k* and have an aridic moisture regime that borders on xeric.

Aridic Duric Haploxerolls are like Typic Haploxerolls except for *d* and *k*, with or without all or any of *b*, *e*, or *h*, and have an aridic moisture regime that borders on xeric.

Calcic Haploxerolls are like Typic Haploxerolls except for *b* or for *b* and *e*.

Calcic Pachic Haploxerolls are like Typic Haploxerolls except for *b* and *c*, with or without *a* or *e*, or both.

Calciorthidic Haploxerolls are like Typic Haploxerolls except for *b* and *k*, with or without *e*, and have an aridic moisture regime.

Cumulic Haploxerolls are like Typic Haploxerolls except for *c* and *f*, with or without all or any of *a*, *b*, *e*, and *k*.

Cumulic Ultic Haploxerolls are like Typic Haploxerolls except for *c*, *f*, and *h*, with or without all or any of *a*, *b*, or *e*.

Entic Haploxerolls are like Typic Haploxerolls except for *e*.

Entic Ultic Haploxerolls are like Typic Haploxerolls except for *e* and *h*.

Fluvaquentic Haploxerolls are like Typic Haploxerolls except for *a* and *f*, with or without *b* or *e*, or both.

Fluventic Haploxerolls are like Typic Haploxerolls except for *f*, with or without *b* or *e*, or both.

Lithic Haploxerolls are like Typic Haploxerolls except for *g*, with or without all or any of *e*, *b*, or *k*.

Lithic Ultic Haploxerolls are like Typic Haploxerolls except for *g* and *h*, with or without *e*.

Pachic Haploxerolls are like Typic Haploxerolls except for *c*, with or without all or any of *a*, *e*, or *k*.

Pachic Ultic Haploxerolls are like Typic Haploxerolls except for *c* and *h*, with or without *a* or *e*, or both.

Torrertic Haploxerolls are like Typic Haploxerolls except for *j*(1) and *k*, with or without all or any of *a*, *b*, *c*, or *f*.

Torrifluventic Haploxerolls are like Typic Haploxerolls except for *f* and *k*, with or without all or any of *b*, *e*, or *h*.

Torriorthentic Haploxerolls are like Typic Haploxerolls except for *e* and *k*, with or without *h*, and they do not have a sandy particle-size class in all subhorizons to a depth of 1 m.

Torripsammentic Haploxerolls are like Typic Haploxerolls except for *e* and *k*, with or without *h*, and they have sandy particle-size class in all subhorizons to a depth of 1 m or more.

Ultic Haploxerolls are like Typic Haploxerolls except for *h*.

Vermic Haploxerolls are like Typic Haploxerolls except for *i*, with or without *b*.

Vertic Haploxerolls are like Typic Haploxerolls except for *j*, with or without all or any or *a*, *b*, *c*, or *f*.

Natrixerolls

Distinctions between Typic Natrixerolls and other subgroups

Typic Natrixerolls are the Natrixerolls that
a. Do not have mottles that have chroma of 2 or less within 75 cm of the soil surface;
b. Do not have a horizon within 1 m of the surface that is >15 cm thick that either contains at least 20 percent durinodes in a nonbrittle matrix or is brittle and has firm consistence when moist; and
c. Have a xeric moisture regime.

Aquic Natrixerolls are like Typic Natrixerolls except for *a*.

Aquic Duric Natrixerolls are like Typic Natrixerolls except for *a* and *b*.

Aridic Natrixerolls are like Typic Natrixerolls except for *c*, and have an aridic moisture regime that borders on xeric.

Duric Natrixerolls are like Typic Natrixerolls except for *b*.

Palexerolls

Distinctions between Typic Palexerolls and other subgroups

Typic Palexerolls are Palexerolls that
a. Have an argillic horizon that has a clayey particle-size class in the upper part and an increase in clay content of 20 percent clay (absolute) or more within a vertical distance of 7.5 cm or of 15 percent clay (absolute) within a distance of 2.5 cm at the upper boundary;
b. Do not have mottles that have chroma of 2 or less within 75 cm of the soil surface;
c. Do not have a petrocalcic horizon that has its upper boundary within 1.5 m of the soil surface;
d. Have a mollic epipedon that is <50 cm thick, or the texture is loamy fine sand or coarser if the mollic epipedon is ≥50 cm thick;
e. Do not have a natric horizon;
f. Have base saturation of >75 percent throughout the argillic horizon or in the upper 50 cm of the argillic horizon, whichever is thinner;
g. Do not have the following combination of characteristics:

(1) Cracks at some period in most years that are 1 cm or more wide at a depth of 50 cm, that are at least 30 cm long in some part, and that extend upward to the surface or to the base of an Ap horizon;

(2) A coefficient of linear extensibility (COLE) of 0.05 or more in a horizon or horizons at least 50 cm thick and a potential linear extensibility of 6 cm or more in the upper 1.5 m of the soil or in the whole soil if a lithic or a paralithic contact is deeper than 50 cm but shallower than 1.5 m; and

(3) More than 35 percent clay in horizons that have a total thickness of >50 cm; and

h. Have a xeric moisture regime.

Aridic Palexerolls are like Typic Palexerolls except for *h* and have an aridic moisture regime that borders on xeric.

Aridic Petrocalcic Palexerolls are like Typic Palexerolls except for *c* and *h*, with or without *a*, and have an aridic moisture regime that borders on xeric.

Pachic Palexerolls are like Typic Palexerolls except for *d*, with or without all or any of *a*, *b*, or *h*.

Petrocalcic Palexerolls are like Typic Palexerolls except for *c*, with or without *a*.

Ultic Palexerolls are like Typic Palexerolls except for *f*, with or without *a*.

Chapter 11

Oxisols

Key To Suborders

CA. Oxisols that are either saturated with water within 30 cm of the mineral soil surface 30 days per year in most years or artificially drained, and have one or more of the following:
1) a histic epipedon;
2) if faintly mottled or not mottled within 50 cm of the soil surface, an epipedon that has a moist color value less than 3.5 and chroma of 2 or less immediately below the epipedon; or
3) if there are distinct or prominent mottles within 50 cm of the soil surface, a chroma of 3 or less or a hue of 2.5Y or yellower in 50 percent or more of the horizon immediately below the epipedon.

Aquox, p. 203

CB. Other Oxisols that have an aridic soil moisture regime.
Torrox, p. 211

CC. Other Oxisols that have an ustic soil moisture regime.
Ustox, p. 219

CD. Other Oxisols that have a perudic soil moisture regime.
Perox, p. 205

CE. Other Oxisols.
Udox, p. 212

Aquox

Key to great groups

CAA. Aquox that have an apparent ECEC of less than 1.50 cmol(+) kg^{-1} clay and a pH value (1 M KCl) of 5 or more in some part of the oxic horizon within a depth of 150 cm of the soil surface.
Acraquox, p. 204

CAB. Other Aquox that have plinthite forming a continuous phase within a depth of 125 cm of the soil surface.
Plinthaquox, p. 205

CAC. Other Aquox that have more than 35 percent base saturation (NH$_4$OAc) in all parts within a depth of 125 cm of the soil surface.
Eutraquox, p. 204

CAD. Other Aquox.
Haplaquox, p. 204

Acraquox

<u>Key to subgroups</u>

CAAA. Acraquox that have More than 5 percent plinthite in some horizon within a depth of 125 cm of the soil surface.

Plinthic Acraquox

CAAB. Other Acraquox that have mottles with chroma of more than 2 in 50 percent or more of the horizon immediately below the epipedon.

Aeric Acraquox

CAAC. Other Acraquox.

Typic Acraquox

Eutraquox

<u>Key to subgroups</u>

CACA. Eutraquox that have a histic epipedon.

Histic Eutraquox

CACB. Other Eutraquox that have more than 5 percent plinthite in some horizon within a depth of 125 cm of the soil surface.

Plinthic Eutraquox

CACC. Other Eutraquox that have mottles with chroma of more than 2 in 50 percent or more of the horizon immediately below the epipedon.

Aeric Eutraquox

CACD. Other Eutraquox that have 16 Kg or more organic carbon per square meter to a depth of one meter, exclusive of surface litter.

Humic Eutraquox

CACE. Other Eutraquox.

Typic Eutraquox

Haplaquox

<u>Key to subgroups</u>

CADA. Haplaquox that have a histic epipedon.

Histic Haplaquox

CADB. Other Haplaquox that have more than 5 percent plinthite in some horizon within a depth of 125 cm of the soil surface.

Plinthic Haplaquox

CADC. Other Haplaquox that have mottles with chroma of more than 2 in 50 percent or more of the horizon immediately below the epipedon.

Aeric Haplaquox

CADD. Other Haplaquox that have 16 Kg or more organic carbon per square meter to a depth of one meter, exclusive of the surface litter.

Humic Haplaquox

CADE. Other Haplaquox.

Typic Haplaquox

Plinthaquox

<u>Key to subgroups</u>

CABA. Plinthaquox that have mottles with chroma of more than 2 in 50 percent or more of the horizon immediately below the epipedon.

Aeric Plinthaquox

CABB. Other Plinthaquox.

Typic Plinthaquox

Perox

Key To Great Groups

CDA: Perox that have a sombric horizon within 150 cm of the soil surface.

Sombriperox, p. 211

CDB: Other Perox that have both an apparent ECEC of less than 1.50 cmol(+) kg^{-1} clay and a pH value (1 M KCl) of 5 or more in some part of the oxic or kandic horizon within a depth of 150 cm of the soil surface.

Acroperox, p. 205

CDC: Other Perox that have more than 35 percent base saturation (NH_4OAc) in all parts within a depth of 125 cm of the soil surface.

Eutroperox, p. 207

CDD: Other Perox that have more than 40 percent clay in the surface 18 cm, after mixing, and the upper boundary of a kandic horizon occurring within a depth of 150 cm of the soil surface.

Kandiperox, p. 209

CDE: Other Perox.

Haploperox, p. 208

Acroperox

<u>Key to subgroups</u>

CDBA. Acroperox that have mottles of 4 or more value moist and 2 or less chroma within a depth of 125 cm of the soil surface and a petroferric contact within 125 cm of the soil surface.

Aquic Petroferric Acroperox

CDBB. Other Acroperox that have a petroferric contact within 125 cm of the soil surface.

Petroferric Acroperox

CDBC. Other Acroperox that have mottles of 4 or more value moist and 2 or less chroma within a depth of 125 cm of the soil surface and a lithic contact within 125 cm of the soil surface.

Aquic Lithic Acroperox

CDBD. Other Acroperox that have a lithic contact within a depth of 125 cm of the soil surface.

Lithic Acroperox

CDBE. Other Acroperox that have a delta pH (KCl pH - 1:1 water pH) with a 0 or net positive charge in some layer 18 cm or more thick within a depth of 125 cm of the soil surface.

Anionic Acroperox

CDBF. Other Acroperox that have more than 5 percent plinthite in some horizon within a depth of 125 cm of the soil surface.

Plinthic Acroperox

CDBG. Other Acroperox that have mottles of 4 or more value moist and 2 or less chroma within a depth of 125 cm of the soil surface.

Aquic Acroperox

CDBH. Other Acroperox that have 16 Kg or more organic carbon per square meter to a depth of one meter, exclusive of surface litter and a color hue of 2.5 YR or redder with moist values of less than 4 in most of the 25 to 125 cm depth from the soil surface.

Humic Rhodic Acroperox

CDBI. Other Acroperox that have 16 Kg or more organic carbon per square meter to a depth of one meter, exclusive of surface litter and color hue of 7.5 YR or yellower with moist values of 6 or more in most of the 25 to 125 cm depth from the soil surface.

Humic Xanthic Acroperox

CDBJ. Other Acroperox that have 16 Kg or more organic carbon per square meter to a depth of one meter, exclusive of surface litter.

Humic Acroperox

CDBK. Other Acroperox that have a color hue of 2.5 YR or redder with moist values of less than 4 in most of the 25 to 125 cm depth from the soil surface.

Rhodic Acroperox

CDBL. Other Acroperox that have a color hue of 7.5 YR or yellower with moist values of 6 or more in most of the 25 to 125 cm depth from the soil surface.

Xanthic Acroperox

CDBM. Other Acroperox.

Typic Acroperox

Eutroperox

<u>Key to subgroups</u>

CDCA. Eutroperox that have mottles of 4 or more value
moist and 2 or less chroma within a depth of 125 cm of the
soil surface and a petroferric contact within 125 cm of the
soil surface.
Aquic Petroferric Eutroperox

CDCB. Other Eutroperox that have a petroferric contact
within 125 cm of the soil surface.
Petroferric Eutroperox

CDCC. Other Eutroperox that have mottles of 4 or more
value moist and 2 or less chroma within a depth of 125 cm
of the soil surface and a lithic contact within 125 cm of the
soil surface.
Aquic Lithic Eutroperox

CDCD. Other Eutroperox that have a lithic contact within a
depth of 125 cm of the soil surface.
Lithic Eutroperox

CDCE. Other Eutroperox that have more than 5 percent
plinthite in some horizon within a depth of 125 cm of the
soil surface and mottles of 4 or more value moist and 2 or
less chroma within a depth of 125 cm of the soil surface.
Plinthaquic Eutroperox

CDCF. Other Eutroperox that have more than 5 percent
plinthite in some horizon within a depth of 125 cm of the
soil surface.
Plinthic Eutroperox

CDCG. Other Eutroperox that have mottles of 4 or more
value moist and 2 or less chroma within a depth of 125 cm
of the soil surface.
Aquic Eutroperox

CDCH. Other Eutroperox that have more than 40 percent
clay in the surface 18 cm after mixing and the upper
boundary of a kandic horizon within a depth or 150 cm of
the soil surface.
Kandiudalfic Eutroperox

CDCI. Other Eutroperox that have 16 Kg or more organic
carbon per square meter to a depth of one meter, exclusive
of surface litter and have the lower boundary of the oxic
horizon within a depth of 125 cm of the soil surface.
Umbreptic Eutroperox

CDCJ. Other Eutroperox that have the lower boundary of
the oxic horizon within a depth of 125 cm of the soil surface.
Inceptic Eutroperox

CDCK. Other Eutroperox that have 16 Kg or more organic
carbon per square meter to a depth of one meter, exclusive
of surface litter and a color hue of 2.5 YR or redder with

moist values of less than 4 in most of the 25 to 125 cm
depth from the soil surface.

Humic Rhodic Eutroperox

CDCL. Other Eutroperox that have 16 Kg or more organic
carbon per square meter to a depth of one meter, exclusive
of surface litter and color hue of 7.5 YR or yellower with
moist values of 6 or more in most of the 25 to 125 cm depth
from the soil surface.

Humic Xanthic Eutroperox

CDCM. Other Eutroperox that have 16 Kg or more organic
carbon per square meter to a depth of one meter, exclusive
of surface litter.

Humic Eutroperox

CDCN. Other Eutroperox that have a color hue of 2.5 YR or
redder with moist values of less than 4 in most of the 25 to
125 cm depth from the soil surface.

Rhodic Eutroperox

CDCO. Other Eutroperox that have a color hue of 7.5 YR or
yellower with moist values of 6 or more in most of
the 25 to 125 cm depth from the soil surface.

Xanthic Eutroperox

CDCP. Other Eutroperox.

Typic Eutroperox

Haploperox

<u>Key to subgroups</u>

CDEA. Haploperox that have mottles of 4 or more value
moist and 2 or less chroma within a depth of 125 cm of the
soil surface and a petroferric contact within 125 cm of the
soil surface.

Aquic Petroferric Haploperox

CDEB. Other Haploperox that have a petroferric contact
within 125 cm of the soil surface.

Petroferric Haploperox

CDEC. Other Haploperox that have mottles of 4 or more
value moist and 2 or less chroma within a depth of 125 cm
of the soil surface and a lithic contact within 125 cm of the
soil surface.

Aquic Lithic Haploperox

CDED. Other Haploperox that have a lithic contact within a
depth of 125 cm of the soil surface.

Lithic Haploperox

CDEE. Other Haploperox that have more than 5 percent
plinthite in some horizon within a depth of 125 cm of the
soil surface and mottles of 4 or more value moist and 2 or
less chroma within a depth of 125 cm of the soil surface.

Plinthaquic Haploperox

CDEF. Other Haploperox that have more than 5 percent plinthite in some horizon within a depth of 125 cm of the soil surface.

Plinthic Haploperox

CDEG. Other Haploperox that have mottles of 4 or more value moist and 2 or less chroma within a depth of 125 cm of the soil surface.

Aquic Haploperox

CDEH. Other Haploperox that have an 18 cm or thicker layer in the upper 75 cm with a bulk density less than 1 g/cc and in which all the Al plus one-half the Fe that is extractable with acid oxalate totals more than 1.0 percent.

Andic Haploperox

CDEI. Other Haploperox that have 16 Kg or more organic carbon per square meter to a depth of one meter, exclusive of surface litter and a color hue of 2.5 YR or redder with moist values of less than 4 in most of the 25 to 125 cm depth from the soil surface.

Humic Rhodic Haploperox

CDEJ. Other Haploperox that have 16 Kg or more organic carbon per square meter to a depth of one meter, exclusive of surface litter and color hue of 7.5 YR or yellower with moist values of 6 or more in most of the 25 to 125 cm depth from the soil surface.

Humic Xanthic Haploperox

CDEK. Other Haploperox that have 16 Kg or more organic carbon per square meter to a depth of one meter, exclusive of surface litter.

Humic Haploperox

CDEL. Other Haploperox that have a color hue of 2.5 YR or redder with moist values of less than 4 in most of the 25 to 125 cm depth from the soil surface.

Rhodic Haploperox

CDEM. Other Haploperox that have a color hue of 7.5 YR or yellower with moist values of 6 or more in most of the 25 to 125 cm depth from the soil surface.

Xanthic Haploperox

CDEN. Other Haploperox.

Typic Haploperox

Kandiperox

<u>Key to subgroups</u>

CDDA. Kandiperox that have mottles of 4 or more value moist and 2 or less chroma within a depth of 125 cm of the soil surface and a petroferric contact within 125 cm of the soil surface.

Aquic Petroferric Kandiperox

CDDB. Other Kandiperox that have a petroferric contact within 125 cm of the soil surface.

Petroferric Kandiperox

CDDC. Other Kandiperox that have mottles of 4 or more value moist and 2 or less chroma within a depth of 125 cm of the soil surface and a lithic contact within 125 cm of the soil surface.

Aquic Lithic Kandiperox

CDDD. Other Kandiperox that have a lithic contact within a depth of 125 cm of the soil surface.

Lithic Kandiperox

CDDE. Other Kandiperox that have more than 5 percent plinthite in some horizon within a depth of 125 cm of the soil surface and mottles of 4 or more value moist and 2 or less chroma within a depth of 125 cm of the soil surface.

Plinthaquic Kandiperox

CDDF. Other Kandiperox that have more than 5 percent plinthite in some horizon within a depth of 125 cm of the soil surface.

Plinthic Kandiperox

CDDG. Other Kandiperox that have mottles of 4 or more value moist and 2 or less chroma within a depth of 125 cm of the soil surface.

Aquic Kandiperox

CDDH. Other Kandiperox that have an 18 cm or thicker layer in the upper 75 cm with a bulk density less than 1 g/cc and in which all the Al plus one-half the Fe that is extractable with acid oxalate totals more than 1.0 percent.

Andic Kandiperox

CDDI. Other Kandiperox that have 16 Kg or more organic carbon per square meter to a depth of one meter, exclusive of surface litter and a color hue of 2.5 YR or redder with moist values of less than 4 in most of the 25 to 125 cm depth from the soil surface.

Humic Rhodic Kandiperox

CDDJ. Other Kandiperox that have 16 Kg or more organic carbon per square meter to a depth of one meter, exclusive of surface litter and color hue of 7.5 YR or yellower with moist values of 6 or more in most of the 25 to 125 cm depth from the soil surface.

Humic Xanthic Kandiperox

CDDK. Other Kandiperox that have 16 Kg or more organic carbon per square meter to a depth of one meter, exclusive of surface litter.

Humic Kandiperox

CDDL. Other Kandiperox that have a color hue of 2.5 YR or redder with moist values of less than 4 in most of the 25 to 125 cm depth from the soil surface.

Rhodic Kandiperox

CDDM. Other Kandiperox that have a color hue of 7.5 YR or yellower with moist values of 6 or more in most of the 25 to 125 cm depth from the soil surface.

Xanthic Kandiperox

CDDN. Other Kandiperox.

Typic Kandiperox

Sombriperox

<u>Key to subgroups</u>

CDAA. Sombriperox that have a petroferric contact within a depth of 125 cm of the soil surface.

Petroferric Sombriperox

CDAB. Other Sombriperox that have a lithic contact within a depth of 125 cm of the soil surface.

Lithic Sombriperox

CDAC. Other Sombriperox that have 16 Kg or more organic carbon per square meter to a depth of one meter, exclusive of surface litter.

Humic Sombriperox

CDAD. Other Sombriperox.

Typic Sombriperox

Torrox

Key To Great Groups

CBA: Torrox that have both an apparent ECEC of less than 1.50 cmol(+) kg^{-1} clay and a pH value (1 M KCl) of 5 or more in some part of the oxic horizon within a depth of 150 cm of the soil surface.

Acrotorrox, p. 211

CBB: Other Torrox that have more than 35 percent base saturation (NH$_4$OAc) in all parts within a depth of 125 cm of the soil surface.

Eutrotorrox, p. 212

CBC: Other Torrox.

Haplotorrox, p. 212

Acrotorrox

<u>Key to subgroups</u>

CBAA. Acrotorrox that have a petroferric contact within a depth of 125 cm of the soil surface.

Petroferric Acrotorrox

CBAB. Other Acrotorrox that have a lithic contact within a depth of 125 cm of the soil surface.

Lithic Acrotorrox

CBAC. Other Acrotorrox.

Typic Acrotorrox

Eutrotorrox

<u>Key to subgroups</u>

CBBA. Eutrotorrox that have a petroferric contact within a depth of 125 cm of the soil surface.

Petroferric Eutrotorrox

CBBB. Other Eutrotorrox that have a lithic contact within a depth of 125 cm of the soil surface.

Lithic Eutrotorrox

CBBC. Other Eutrotorrox.

Typic Eutrotorrox

Haplotorrox

<u>Key to subgroups</u>

CBCA. Haplotorrox that have a petroferric contact within a depth of 125 cm of the soil surface.

Petroferric Haplotorrox

CBCB. Other Haplotorrox that have a lithic contact within a depth of 125 cm of the soil surface.

Lithic Haplotorrox

CBCC. Other Haplotorrox.

Typic Haplotorrox

Udox

Key to great groups

CEA: Udox that have a sombric horizon within a depth of 150 cm of the soil surface.

Sombriudox, p. 218

CEB: Other Udox that have both an apparent ECEC of less than 1.50 cmol(+) kg^{-1} clay and a pH value (1 M KCl) of 5 or more in some part of the oxic or kandic horizon within a depth of 150 cm of the soil surface.

Acrudox, p. 213

CEC: Other Udox that have more than 35 percent base saturation (NH$_4$OAc) in all parts within a depth of 125 cm of the soil surface.

Eutrudox, p. 214

CED: Other Udox that have more than 40 percent clay in the surface 18 cm after mixing, and the upper boundary of a kandic horizon occurring within a depth of 150 cm of the surface.

Kandiudox, p. 217

CEE: Other Udox.

Hapludox, p. 215

Acrudox

<u>Key to subgroups</u>

CEBA. Acrudox that have mottles of 4 or more value moist and 2 or less chroma within a depth of 125 cm of the soil surface and a petroferric contact within 125 cm of the soil surface.

Aquic Petroferric Acrudox

CEBB. Other Acrudox that have a petroferric contact within 125 cm of the soil surface.

Petroferric Acrudox

CEBC. Other Acrudox that have mottles of 4 or more value moist and 2 or less chroma within a depth of 125 cm of the soil surface and a lithic contact within 125 cm of the soil surface.

Aquic Lithic Acrudox

CEBD. Other Acrudox that have a lithic contact within a depth of 125 cm of the soil surface.

Lithic Acrudox

CEBE. Other Acrudox that have mottles of 4 or more value moist and 2 or less chroma within a depth of 125 cm of the soil surface and have a delta pH (KCl pH - 1:1 water pH) with a 0 or net positive charge in some layer 18 cm or more thick within a depth of 125 cm of the soil surface.

Aquic Anionic Acrudox

CEBF. Other Acrudox that have a delta pH (KCl pH - 1:1 water pH) with a 0 or net positive charge in some layer 18 cm or more thick within a depth of 125 cm of the soil surface.

Anionic Acrudox

CEBG. Other Acrudox that have more than 5 percent plinthite in some horizon within a depth of 125 cm of the soil surface.

Plinthic Acrudox

CEBH. Other Acrudox that have mottles of 4 or more value moist and 2 or less chroma within a depth of 125 cm of the soil surface.

Aquic Acrudox

CEBI. Other Acrudox that have more than 35 percent base saturation (NH_4OAc) in all parts within a depth of 125 cm of the soil surface.

Eutric Acrudox

CEBJ. Other Acrudox that have 16 Kg or more organic carbon per square meter to a depth of one meter, exclusive of surface litter and a color hue of 2.5 YR or redder with moist values of less than 4 in most of the 25 to 125 cm depth from the soil surface.

Humic Rhodic Acrudox

CEBK. Other Acrudox that have 16 Kg or more organic carbon per square meter to a depth of one meter, exclusive of surface litter and color hue of 7.5 YR or yellower with moist values of 6 or more in most of the 25 to 125 cm depth from the soil surface.

Humic Xanthic Acrudox

CEBL. Other Acrudox that have 16 Kg or more organic carbon per square meter to a depth of one meter, exclusive of surface litter.

Humic Acrudox

CEBM. Other Acrudox that have a color hue of 2.5 YR or redder with moist values of less than 4 in most of the 25 to 125 cm depth from the soil surface.

Rhodic Acrudox

CEBN. Other Acrudox that have a color hue of 7.5 YR or yellower with moist values of 6 or more in most of the 25 to 125 cm depth from the soil surface.

Xanthic Acrudox

CEBO. Other Acrudox.

Typic Acrudox

Eutrudox

<u>Key to subgroups</u>

CECA. Eutrudox that have mottles of 4 or more value moist and 2 or less chroma within a depth of 125 cm of the soil surface and a petroferric contact within 125 cm of the soil surface.

Aquic Petroferric Eutrudox

CECB. Other Eutrudox that have a petroferric contact within 125 cm of the soil surface.

Petroferric Eutrudox

CECC. Other Eutrudox that have mottles of 4 or more value moist and 2 or less chroma within a depth of 125 cm of the soil surface and a lithic contact within 125 cm of the soil surface.

Aquic Lithic Eutrudox

CECD. Other Eutrudox that have a lithic contact within a depth of 125 cm of the soil surface.

Lithic Eutrudox

CECE. Other Eutrudox that have more than 5 percent plinthite in some horizon within a depth of 125 cm of the soil surface and mottles of 4 or more value moist and 2 or less chroma within a depth of 125 cm of the soil surface.

Plinthaquic Eutrudox

CECF. Other Eutrudox that have more than 5 percent plinthite in some horizon within a depth of 125 cm of the soil surface.

Plinthic Eutrudox

CECG. Other Eutrudox that have mottles of 4 or more value moist and 2 or less chroma within a depth of 125 cm of the soil surface.

Aquic Eutrudox

CECH. Other Eutrudox that have more than 40 percent clay in the surface 18 cm after mixing and the upper boundary of a kandic horizon within a depth or 150 cm of the soil surface.

Kandiudalfic Eutrudox

CECI. Other Eutrudox that have 16 Kg or more organic carbon per square meter to a depth of one meter, exclusive of surface litter and have the lower boundary of the oxic horizon within a depth of 125 cm of the soil surface.

Umbreptic Eutrudox

CECJ. Other Eutrudox that have the lower boundary of the oxic horizon within a depth of 125 cm of the soil surface.

Inceptic Eutrudox

CECK. Other Eutrudox that have 16 Kg or more organic carbon per square meter to a depth of one meter, exclusive of surface litter and a color hue of 2.5 YR or redder with moist values of less than 4 in most of the 25 to 125 cm depth from the soil surface.

Humic Rhodic Eutrudox

CECL. Other Eutrudox that have 16 Kg or more organic carbon per square meter to a depth of one meter, exclusive of surface litter and color hue of 7.5 YR or yellower with moist values of 6 or more in most of the 25 to 125 cm depth from the soil surface.

Humic Xanthic Eutrudox

CECM. Other Eutrudox that have 16 Kg or more organic carbon per square meter to a depth of one meter, exclusive of surface litter.

Humic Eutrudox

CECN. Other Eutrudox that have a color hue of 2.5 YR or redder with moist values of less than 4 in most of the 25 to 125 cm depth from the soil surface.

Rhodic Eutrudox

CECO. Other Eutrudox that have a color hue of 7.5 YR or yellower with moist values of 6 or more in most of the 25 to 125 cm depth from the soil surface.

Xanthic Eutrudox

CECP. Other Eutrudox.

Typic Eutrudox

Hapludox

<u>Key to subgroups</u>

CEEA. Hapludox that have mottles of 4 or more value moist and 2 or less chroma within a depth of 125 cm of the soil surface and a petroferric contact within 125 cm of the soil surface.

Aquic Petroferric Hapludox

CEEB. Other Hapludox that have a petroferric contact within 125 cm of the soil surface.

Petroferric Hapludox

CEEC. Other Hapludox that have mottles of 4 or more value moist and 2 or less chroma within a depth of 125 cm of the soil surface and a lithic contact within 125 cm of the soil surface.

Aquic Lithic Hapludox

CEED. Other Hapludox that have a lithic contact within a depth of 125 cm of the soil surface.

Lithic Hapludox

CEEE. Other Hapludox that have more than 5 percent plinthite in some horizon within a depth of 125 cm of the soil surface and mottles of 4 or more value moist and 2 or less chroma within a depth of 125 cm of the soil surface.

Plinthaquic Hapludox

CEEF. Other Hapludox that have more than 5 percent plinthite in some horizon within a depth of 125 cm of the soil surface.

Plinthic Hapludox

CEEG. Other Hapludox that have mottles of 4 or more value moist and 2 or less chroma within a depth of 125 cm of the soil surface.

Aquic Hapludox

CEEH. Other Hapludox that have the lower boundary of the oxic horizon within a depth or 125 cm of the soil surface.

Inceptic Hapludox

CEEI. Other Hapludox that have an 18 cm or thicker layer in the upper 75 cm with a bulk density less than 1 g/cc and in which all the Al plus one-half the Fe that is extractable with acid oxalate totals more than 1.0 percent.

Andic Hapludox

CEEJ. Other Hapludox that have 16 Kg or more organic carbon per square meter to a depth of one meter, exclusive of surface litter and a color hue of 2.5 YR or redder with moist values of less than 4 in most of the 125 to 125 cm depth from the soil surface.

Humic Rhodic Hapludox

CEEK. Other Hapludox that have 16 Kg or more organic carbon per square meter to a depth of one meter, exclusive of surface litter and color hue of 7.5 YR or yellower with moist values of 6 or more in most of the 25 to 125 cm depth from the soil surface.

Humic Xanthic Hapludox

CEEL. Other Hapludox that have 16 Kg or more organic carbon per square meter to a depth of one meter, exclusive of surface litter.

Humic Hapludox

CEEM. Other Hapludox that have a color hue of 2.5 YR or redder with moist values of less than 4 in most of the 25 to 125 cm depth from the soil surface.

Rhodic Hapludox

CEEN. Other Hapludox that have a color hue of 7.5 YR or yellower with moist values of 6 or more in most of the 25 to 125 cm depth from the soil surface.

Xanthic Hapludox

CEEO. Other Hapludox.

Typic Hapludox

Kandiudox

<u>Key to subgroups</u>

CEDA. Kandiudox that have mottles of 4 or more value moist and 2 or less chroma within a depth of 125 cm of the soil surface and a petroferric contact within 125 cm of the soil surface.

Aquic Petroferric Kandiudox

CEDB. Other Kandiudox that have a petroferric contact within 125 cm of the soil surface.

Petroferric Kandiudox

CEDC. Other Kandiudox that have mottles of 4 or more value moist and 2 or less chroma within a depth of 125 cm of the soil surface and a lithic contact within 125 cm of the soil surface.

Aquic Lithic Kandiudox

CEDD. Other Kandiudox that have a lithic contact within a depth of 125 cm of the soil surface.

Lithic Kandiudox

CEDE. Other Kandiudox that have more than 5 percent plinthite in some horizon within a depth of 125 cm of the soil surface and mottles of 4 or more value moist and 2 or less chroma within a depth of 125 cm of the soil surface.

Plinthaquic Kandiudox

CEDF. Other Kandiudox that have more than 5 percent plinthite in some horizon within a depth of 125 cm of the soil surface.

Plinthic Kandiudox

CEDG. Other Kandiudox that have mottles of 4 or more value moist and 2 or less chroma within a depth of 125 cm of the soil surface.

Aquic Kandiudox

CEDH. Other Kandiudox that have an 18 cm or thicker layer in the upper 75 cm with a bulk density less than 1 g/cc and in which all the Al plus one-half the Fe that is extractable with acid oxalate totals more than 1.0 percent.

Andic Kandiudox

CEDI. Other Kandiudox that have 16 Kg or more organic carbon per square meter to a depth of one meter, exclusive of surface litter and a color hue of 2.5 YR or redder with moist values of less than 4 in most of the 25 to 125 cm depth from the soil surface.

Humic Rhodic Kandiudox

CEDJ. Other Kandiudox that have 16 Kg or more organic carbon per square meter to a depth of one meter, exclusive of surface litter and color hue of 7.5 YR or yellower with moist values of 6 or more in most of the 25 to 125 cm depth from the soil surface.

Humic Xanthic Kandiudox

CEDK. Other Kandiudox that have 16 Kg or more organic carbon per square meter to a depth of one meter, exclusive of surface litter.

Humic Kandiudox

CEDL. Other Kandiudox that have a color hue of 2.5 YR or redder with moist values of less than 4 in most of the 25 to 125 cm depth from the soil surface.

Rhodic Kandiudox

CEDM. Other Kandiudox that have a color hue of 7.5 YR or yellower with moist values of 6 or more in most of the 25 to 125 cm depth from the soil surface.

Xanthic Kandiudox

CEDN. Other Kandiudox.

Typic Kandiudox

Sombriudox

<u>Key to subgroups</u>

CEAA. Sombriudox that have a petroferric contact within a depth of 125 cm of the soil surface.

Petroferric Sombriudox

CEAB. Other Sombriudox that have a lithic contact within a depth of 125 cm of the soil surface.

Lithic Sombriudox

CEAC. Other Sombriudox that have 16 Kg or more organic carbon per square meter to a depth of one meter, exclusive of surface litter.

Humic Sombriudox

CEAD. Other Sombriudox.

Typic Sombriudox

Ustox

Key to great groups

CCA: Ustox that have a sombric horizon within 150 cm of the soil surface.

Sombriustox, p. 224

CCB: Other Ustox that have both an apparent ECEC of less than 1.50 cmol(+) kg^{-1} clay and a pH value (1 M KCl) of 5 or more in some part of the oxic or kandic horizon within a depth of 150 cm of the soil surface.

Acrustox, p. 219

CCC: Other Ustox that have more than 35 percent base saturation (NH$_4$OAc) in all parts within a depth of 125 cm of the soil surface.

Eutrustox, p. 220

CCD: Other Ustox that have more than 40 percent clay in the surface 18 cm after mixing, and the upper boundary of a kandic horizon occurring within a depth of 150 cm of the soil surface.

Kandiustox, p. 223

CCE: Other Ustox.

Haplustox, p. 222

Acrustox

<u>Key to subgroups</u>

CCBA. Acrustox that have mottles of 4 or more value moist and 2 or less chroma within a depth of 125 cm of the soil surface and a petroferric contact within 125 cm of the soil surface.

Aquic Petroferric Acrustox

CCBB. Other Acrustox that have a petroferric contact within 125 cm of the soil surface.

Petroferric Acrustox

CCBC. Other Acrustox that have mottles of 4 or more value moist and 2 or less chroma within a depth of 125 cm of the soil surface and a lithic contact within 125 cm of the soil surface.

Aquic Lithic Acrustox

CCBD. Other Acrustox that have a lithic contact within a depth of 125 cm of the soil surface.

Lithic Acrustox

CCBE. Other Acrustox that have mottles of 4 or more value moist and 2 or less chroma within a depth of 125 cm of the soil surface and have a delta pH (KCl pH - 1:1 water pH) with a 0 or net positive charge in some layer 18 cm or more thick within a depth of 125 cm of the soil surface.

Aquic Anionic Acrustox

CCBF. Other Acrustox that have a delta pH (KCl pH - 1:1 water pH) with a 0 or net positive charge in some layer 18 cm or more thick within a depth of 125 cm of the soil surface.

Anionic Acrustox

CCBG. Other Acrustox that have more than 5 percent plinthite in some horizon within a depth of 125 cm of the soil surface.

Plinthic Acrustox

CCBH. Other Acrustox that have mottles of 4 or more value moist and 2 or less chroma within a depth of 125 cm of the soil surface.

Aquic Acrustox

CCBI. Other Acrustox that have more than 35 percent base saturation (NH$_4$OAc) in all parts within a depth of 125 cm of the soil surface.

Eutric Acrustox

CCBJ. Other Acrustox that have 16 Kg or more organic carbon per square meter to a depth of one meter, exclusive of surface litter and a color hue of 2.5 YR or redder with moist values of less than 4 in most of the 25 to 125 cm depth from the soil surface.

Humic Rhodic Acrustox

CCBK. Other Acrustox that have 16 Kg or more organic carbon per square meter to a depth of one meter, exclusive of surface litter and color hue of 7.5 YR or yellower with moist values of 6 or more in most of the 25 to 125 cm depth from the soil surface.

Humic Xanthic Acrustox

CCBL. Other Acrustox that have 16 Kg or more organic carbon per square meter to a depth of one meter, exclusive of surface litter.

Humic Acrustox

CCBM. Other Acrustox that have a color hue of 2.5 YR or redder with moist values of less than 4 in most of the 25 to 125 cm depth from the soil surface.

Rhodic Acrustox

CCBN. Other Acrustox that have a color hue of 7.5 YR or yellower with moist values of 6 or more in most of the 25 to 125 cm depth from the soil surface.

Xanthic Acrustox

CCBO. Other Acrustox.

Typic Acrustox

Eutrustox

Key to subgroups

CCCA. Eutrustox that have mottles of 4 or more value moist and 2 or less chroma within a depth of 125 cm of the soil surface and a petroferric contact within 125 cm of the soil surface.

Aquic Petroferric Eutrustox

CCCB. Other Eutrustox that have a petroferric contact within 125 cm of the soil surface.

Petroferric Eutrustox

CCCC. Other Eutrustox that have mottles of 4 or more value moist and 2 or less chroma within a depth of 125 cm of the soil surface and a lithic contact within 125 cm of the soil surface.

Aquic Lithic Eutrustox

CCCD. Other Eutrustox that have a lithic contact within a depth of 125 cm of the soil surface.

Lithic Eutrustox

CCCE. Other Eutrustox that have more than 5 percent plinthite in some horizon within a depth of 125 cm of the soil surface and mottles of 4 or more value moist and 2 or less chroma within a depth of 125 cm of the soil surface.

Plinthaquic Eutrustox

CCCF. Other Eutrustox that have more than 5 percent plinthite in some horizon within a depth of 125 cm of the soil surface.

Plinthic Eutrustox

CCCG. Other Eutrustox that have mottles of 4 or more value moist and 2 or less chroma within a depth of 125 cm of the soil surface.

Aquic Eutrustox

CCCH. Other Eutrustox that have more than 40 percent clay in the surface 18 cm after mixing and the upper boundary of a kandic horizon within a depth or 150 cm of the soil surface.

Kandiustalfic Eutrustox

CCCI. Other Eutrustox that have 16 Kg or more organic carbon per square meter to a depth of one meter, exclusive of surface litter and have the lower boundary of the oxic horizon within a depth of 125 cm of the soil surface.

Umbreptic Eutrustox

CCCJ. Other Eutrustox that have the lower boundary of the oxic horizon within a depth of 125 cm of the soil surface.

Inceptic Eutrustox

CCCK. Other Eutrustox that have 16 Kg or more organic carbon per square meter to a depth of one meter, exclusive of surface litter and a color hue of 2.5 YR or redder with moist values of less than 4 in most of the 25 to 125 cm depth from the soil surface.

Humic Rhodic Eutrustox

CCCL. Other Eutrustox that have 16 Kg or more organic carbon per square meter to a depth of one meter, exclusive of surface litter and color hue of 7.5 YR or yellower with moist values of 6 or more in most of the 25 to 125 cm depth from the soil surface.

Humic Xanthic Eutrustox

CCCM. Other Eutrustox that have 16 Kg or more organic carbon per square meter to a depth of one meter, exclusive of surface litter.

Humic Eutrustox

CCCN. Other Eutrustox that have a color hue of 2.5 YR or redder with moist values of less than 4 in most of the 25 to 125 cm depth from the soil surface.

Rhodic Eutrustox

CCCO. Other Eutrustox that have a color hue of 7.5 YR or yellower with moist values of 6 or more in most of the 25 to 125 cm depth from the soil surface.

Xanthic Eutrustox

CCCP. Other Eutrustox.

Typic Eutrustox

Haplustox

<u>Key to subgroups</u>

CCEA. Haplustox that have mottles of 4 or more value moist and 2 or less chroma within a depth of 125 cm of the soil surface and a petroferric contact within 125 cm of the soil surface.

Aquic Petroferric Haplustox

CCEB. Other Haplustox that have a petroferric contact within 125 cm of the soil surface.

Petroferric Haplustox

CCEC. Other Haplustox that have mottles of 4 or more value moist and 2 or less chroma within a depth of 125 cm of the soil surface and a lithic contact within 125 cm of the soil surface.

Aquic Lithic Haplustox

CCED. Other Haplustox that have a lithic contact within a depth of 125 cm of the soil surface.

Lithic Haplustox

CCEE. Other Haplustox that have more than 5 percent plinthite in some horizon within a depth of 125 cm of the soil surface and mottles of 4 or more value moist and 2 or less chroma within a depth of 125 cm of the soil surface.

Plinthaquic Haplustox

CCEF. Other Haplustox that have more than 5 percent plinthite in some horizon within a depth of 125 cm of the soil surface.

Plinthic Haplustox

CCEG. Other Haplustox that have mottles of 4 or more value moist and 2 or less chroma within a depth of 125 cm of the soil surface and have the lower boundary of the oxic horizon within a depth or 125 cm of the soil surface.

Aqueptic Haplustox

CCEH. Other Haplustox that have mottles of 4 or more value moist and 2 or less chroma within a depth of 125 cm of the soil surface.

Aquic Haplustox

CCEI. Other Haplustox that have the lower boundary of the oxic horizon within a depth or 125 cm of the soil surface.

Inceptic Haplustox

CCEJ. Other Haplustox that have 16 Kg or more organic carbon per square meter to a depth of one meter, exclusive of surface litter and a color hue of 2.5 YR or redder with moist values of less than 4 in most of the 25 to 125 cm depth from the soil surface.

Humic Rhodic Haplustox

CCEK. Other Haplustox that have 16 Kg or more organic carbon per square meter to a depth of one meter, exclusive of surface litter and color hue of 7.5 YR or yellower with moist values of 6 or more in most of the 25 to 125 cm depth from the soil surface.

Humic Xanthic Haplustox

CCEL. Other Haplustox that have 16 Kg or more organic carbon per square meter to a depth of one meter, exclusive of surface litter.

Humic Haplustox

CCEM. Other Haplustox that have a color hue of 2.5 YR or redder with moist values of less than 4 in most of the 25 to 125 cm depth from the soil surface.

Rhodic Haplustox

CCEN. Other Haplustox that have a color hue of 7.5 YR or yellower with moist values of 6 or more in most of the 25 to 125 cm depth from the soil surface.

Xanthic Haplustox

CCEO. Other Haplustox.

Typic Haplustox

Kandiustox

<u>Key to subgroups</u>

CCDA. Kandiustox that have mottles of 4 or more value moist and 2 or less chroma within a depth of 125 cm of the soil surface and a petroferric contact within 125 cm of the soil surface.

Aquic Petroferric Kandiustox

CCDB. Other Kandiustox that have a petroferric contact within 125 cm of the soil surface.

Petroferric Kandiustox

CCDC. Other Kandiustox that have mottles of 4 or more value moist and 2 or less chroma within a depth of 125 cm of the soil surface and a lithic contact within 125 cm of the soil surface.

Aquic Lithic Kandiustox

CCDD. Other Kandiustox that have a lithic contact within a depth of 125 cm of the soil surface.

Lithic Kandiustox

CCDE. Other Kandiustox that have more than 5 percent plinthite in some horizon within a depth of 125 cm of the soil surface and mottles of 4 or more value moist and 2 or less chroma within a depth of 125 cm of the soil surface.

Plinthaquic Kandiustox

CCDF. Other Kandiustox that have more than 5 percent plinthite in some horizon within a depth of 125 cm of the soil surface.

Plinthic Kandiustox

CCDG. Other Kandiustox that have mottles of 4 or more value moist and 2 or less chroma within a depth of 125 cm of the soil surface.

Aquic Kandiustox

CCDH. Other Kandiustox that have 16 Kg or more organic carbon per square meter to a depth of one meter, exclusive of surface litter and a color hue of 2.5 YR or redder with moist values of less than 4 in most of the 25 to 125 cm depth from the soil surface.

Humic Rhodic Kandiustox

CCDI. Other Kandiustox that have 16 Kg or more organic carbon per square meter to a depth of one meter, exclusive of surface litter and color hue of 7.5 YR or yellower with moist values of 6 or more in most of the 25 to 125 cm depth from the soil surface.

Humic Xanthic Kandiustox

CCDJ. Other Kandiustox that have 16 Kg or more organic carbon per square meter to a depth of one meter, exclusive of surface litter.

Humic Kandiustox

CCDK. Other Kandiustox that have a color hue of 2.5 YR or redder with moist values of less than 4 in most of the 25 to 125 cm depth from the soil surface.

Rhodic Kandiustox

CCDL. Other Kandiustox that have a color hue of 7.5 YR or yellower with moist values of 6 or more in most of the 25 to 125 cm depth from the soil surface.

Xanthic Kandiustox

CCDM. Other Kandiustox.

Typic Kandiustox

Sombriustox

<u>Key to subgroups</u>

CCAA. Sombriustox that have a petroferric contact within a depth of 125 cm of the soil surface.

Petroferric Sombriustox

CCAB. Other Sombriustox that have a lithic contact within a depth of 125 cm of the soil surface.

Lithic Sombriustox

CCAC. Other Sombriustox that have 16 Kg or more organic carbon per square meter to a depth of one meter, exclusive of surface litter.

Humic Sombriustox

CCAD. Other Sombriustox.

Typic Sombriustox

OXI

Chapter 12

Spodosols

Key To Suborders

BA. Spodosols that either have an aquic moisture regime[1] or are artificially drained and have characteristics associated with wetness, namely one or more of the following:
1. A histic epipedon;
2. Mottling in an albic horizon or in the upper part of the spodic horizon;
3. A duripan in the albic horizon;
4. If free iron and manganese are absent or if the color value, moist, is less than 4 in the upper part of the spodic horizon, either
    a. Have any color if there are no coatings of iron oxides on the individual grains of silt and sand in or immediately below the spodic horizon wherever the value, moist, is 4 or more; or
    b. Have fine or medium mottles of iron or manganese in the materials immediately below the spodic horizon;
5. A placic horizon that rests on a fragipan or on a spodic horizon or on an albic horizon that is underlain by a spodic horizon but is not in a spodic horizon.

Aquods, p. 227

BB. Other Spodosols that have a spodic horizon in which the ratio of free iron (by dithionite-citrate) to carbon (both elemental) is 6 or more in all subhorizons.

Ferrods, p. 231

BC. Other Spodosols that have a spodic horizon in which some subhorizon that is present in more than half of each pedon has a ratio of free iron to carbon of <0.2.

Humods, p. 231

BD. Other Spodosols.

Orthods, p. 233

Aquods

Key to great groups

BAA. Aquods that have a fragipan below the spodic horizon but do not have a placic horizon above the fragipan.

Fragiaquods, p. 228

BAB. Other Aquods that do not have a placic horizon but have a cryic temperature regime.

Cryaquods, p. 228

BAC. Other Aquods that have a strongly cemented or indurated albic horizon that does not slake in water when a dry fragment is immersed.

Duraquods, p. 228

---

[1]If a placic horizon, duripan, or fragipan is present, the soil need not be saturated below that horizon.

BAD. Other Aquods that have a placic horizon that rests on a spodic horizon or on a fragipan or on an albic horizon that rests on a spodic horizon.

Placaquods, p. 230

BAE. Other Aquods that have a mean annual soil temperature of 8°C or higher and mean summer and mean winter soil temperatures at a depth of 50 cm that differ by <5°C.

Tropaquods, p. 230

BAF. Other Aquods that have in >50 percent of each pedon a spodic horizon in which some subhorizon has a ratio of free iron (by dithionite-citrate) to carbon (both elemental) that is <0.2.

Haplaquods, p. 229

BAG. Other Aquods.

Sideraquods, p. 230

Cryaquods

Distinctions between Typic Cryaquods and other subgroups

Typic Cryaquods are the Cryaquods that
a. Do not have a lithic contact within 50 cm of the surface of the mineral soil;
b. Have a mean annual soil temperature higher than 0°C;
c. Have a ratio of free iron to carbon (elemental) of <0.2 in some subhorizon;
d. Do not have mottles above the spodic horizon that are due to segregation of iron;
e. Do not have an argillic horizon underlying the spodic horizon; and
f. Have a continuous spodic horizon that is 10 cm or more thick or that is very firm when moist.
Lithic Cryaquods are like Typic Cryaquods except for *a* or for *a* and *b*.
Pergelic Cryaquods are like Typic Cryaquods except for *b*.
Pergelic Sideric Cryaquods are like Typic Cryaquods except for *b* and *c* or for *b*, *c*, and *d*.
Sideric Cryaquods are like Typic Cryaquods except for *c* or for *c* and *d*.

Duraquods

Duraquods are the Aquods that have a duripan in the albic horizon and have a temperature regime warmer than that of Cryaquods.

Fragiaquods

Distinctions between Typic Fragiaquods and other subgroups

Typic Fragiaquods are the Fragiaquods that
a. Have a frigid or warmer temperature regime;
b. Do not have a histic epipedon;
c. Have <5 percent by volume of iron-cemented nodules 2.5 to 30 cm in diameter in any subhorizon of the spodic horizon;
d. Do not have a surface horizon >30 cm thick that meets all requirements of a plaggen epipedon except thickness;

e. Do not have an intermittent upper black subhorizon of the spodic horizon that has a ratio of free iron (elemental) to carbon that is <0.2; or if plowed and the Ap horizon rests directly on the spodic horizon, do not have tongues of such a subhorizon;
f. Do not have an argillic or kandic horizon.
Alfic Fragiaquods are like Typic Fragiaquods except for *f*.
Cryic Fragiaquods are like Typic Fragiaquods except for *a* or for *a* and *b*.

Haplaquods

Distinctions between Typic Haplaquods and other subgroups

Typic Haplaquods are the Haplaquods that
a. Have an umbric epipedon or one that would meet the requirements for an umbric epipedon if it were plowed to a depth of 25 to 30 cm;
b. Do not have an argillic or kandic horizon underlying the spodic horizon;
c. Do not have a sandy epipedon (loamy fine sand or coarser throughout) that is 75 cm or more thick;
d. Have a spodic horizon that has a weighted average of 0.6 percent or more organic carbon in the matrix of the upper 30 cm of the spodic horizon or have an upper subhorizon of the spodic horizon that
    (1) Has 2.3 percent or more organic carbon in the upper 2 cm; and
    (2) Is continuous or is present in more than 90 percent of each pedon;
e. Have <5 percent by volume of iron-cemented nodules 2.5 to 30 cm in diameter in any subhorizon of the spodic horizon;
f. Do not have a lithic contact within 50 cm of the mineral soil surface;
g. Do not have a surface horizon >30 cm thick that meets all requirements of a plaggen epipedon except thickness;
h. Do not have a histic epipedon; and
i. Do not have a placic horizon in or below the spodic horizon.
Aeric Haplaquods are like Typic Haplaquods except for *a*.
Alfic Haplaquods are like TYpic Haplaquods except for *b*, with or without *a* or *d*, or both, and have either base saturation of 35 percent or more (by sum of cations) in some part of the argillic or kandic horizon or have a mean annual soil temperature lower than 8°C.
Alfic Arenic Haplaquods are like Typic Haplaquods except for *a*, *b*, and *c*, with or without c, with or without d, and have base saturation of 35 percent or more (by sum of cations) in some part of the argillic horizon or kandic or have a mean annual soil temperature lower than 8°C and the upper boundary of the spodic horizon is between 75 cm and 1.25 m below the soil surface.
Arenic Haplaquods are like Typic Haplaquods except for *a* and *c*, and the upper boundary of the spodic horizon is between 75 cm and 1.25 m below the soil surface.
Arenic Ultic Haplaquods are like Typic Haplaquods except for *a*, *b*, and *c* with or without d and have base saturation (by sum of cations) of less than 35 percent throughout the Argillic or kandic horizon and have a mean annual soil temperature of 8°C or higher and the upper boundary of the

spodic horizon is between 75 cm and 1.25 m below the soil surface.

Entic Haplaquods are like Typic Haplaquods except for *a* and *d*.

Ferrudalfic Haplaquods are like Typic Haplaquods except for *b* and *e*, with or without *a* or *d*, or both.

Grossarenic Haplaquods are like Typic Haplaquods except for *a* and *c*, and the upper boundary of the spodic horizon is between 1.25 m and 2 m below the soil surface.

Lithic Haplaquods are like Typic Haplaquods except for *f*, with or without *a* or *d*, or both.

Placic Haplaquods are like Typic Haplaquods except for *i*, with or without *a* or *h*, or both.

Ultic Haplaquods are like Typic Haplaquods except for *b*, with or without *a* or *d*, or both, and have base saturation (by sum of cations) <35 percent throughout the argillic or kandic horizon, and have a mean annual soil temperature of 8°C or higher.

Placaquods

Placaquods are the Aquods that have a placic horizon that rests on a spodic horizon, on a fragipan, or on an albic horizon that is underlain by a fragipan. There may be a histic epipedon at the surface. The horizons above the placic horizon are saturated with water at some period and have faint to distinct mottles of low chroma.

Sideraquods

Distinctions between Typic Sideraquods and other subgroups

Typic Sideraquods are the Sideraquods that
a. Have a spodic horizon that either
   (1) Is at least very firm in some subhorizon when moist; or
   (2) Is >10 cm thick and contains 1.2 percent or more organic carbon in the upper 10 cm;
b. Do not have a histic epipedon;
c. Do not have an argillic or kandic horizon.

Alfic Sideraquods are like Typic Sideraquods except for *c*, with or without *a*, and they have base saturation of 35 percent or more (by sum of cations) in some part of the argillic or kandic horizon or they have a mean annual soil temperature lower than 8°C.

Entic Sideraquods are like Typic Sideraquods except for *a*.

Tropaquods

Distinctions between Typic Tropaquods and other subgroups

Typic Tropaquods are the Tropaquods that
a. Do not have a histic epipedon;
b. Have <5 percent by volume of iron-cemented nodules, 2.5 to 30 cm in diameter, in any subhorizon of the spodic horizon;
c. Do not have an argillic or kandic horizon underlying the spodic horizon;
d. Do not have a lithic contact within 50 cm of the surface;

e. Have an umbric epipedon or one that would meet the requirements for an umbric epipedon if plowed to a depth of 25 to 30 cm;

f. Do not have a sandy epipedon (loamy fine sand or coarser throughout) that is 75 cm or more thick on the average for the thinnest half of the pedon;

g. Have a spodic horizon that has a weighted average of 0.6 percent or more organic carbon in the matrix of the upper 30 cm of the spodic horizon or have an upper subhorizon of the spodic horizon that

    (1) Has 2.3 percent or more organic carbon in the upper 2 cm; and

    (2) Is continuous or is present in >90 percent of each pedon; and

h. Have in 50 percent or more of each pedon a spodic horizon in which some subhorizon has a ratio of free iron (by dithionite-citrate) to carbon (both elemental) of <0.2.

Aeric Tropaquods are like Typic Tropaquods except for *e*.

Aeric Arenic Tropaquods are like Typic Tropaquods except for *e* and *f*, and the upper boundary of the spodic horizon is between 75 cm and 1.25 m below the soil surface.

Aeric Grossarenic Tropaquods are like Typic Tropaquods except for *e* and *f*, and the upper boundary of the spodic horizon is between 1.25 m and 2 m below the soil surface.

Entic Tropaquods are like Typic Tropaquods except for *g*.

Histic Tropaquods are like Typic Tropaquods except for *a*.

Histic Lithic Tropaquods are like Typic Tropaquods except for *a* and *d*.

Ultic Tropaquods are like Typic Tropaquods except for *c*, with or without *e*.

Ferrods

This suborder is provisional. Ferrods are not known to occur in the United States, but the suborder is provided for use elsewhere. The classification has not been developed.

Ferrods are the Spodosols that

1. Have a spodic horizon that has in all subhorizons a ratio of percentage of free iron (by dithionite-citrate) to percentage of carbon (both elemental) of 6 or more; and

2. Do not have an aquic moisture regime or artificial drainage or do not have the characteristics associated with wetness as defined for Aquods.

Humods

Key to great groups

BCA. Humods that have a placic horizon in the spodic horizon.

                                                  Placohumods, p. 233

BCB. Other Humods that have an isomesic or warmer iso temperature regime.

                                                  Tropohumods, p. 233

BCC. Other Humods that have a fragipan below the spodic horizon.

                                                  Fragihumods, p. 232

BCD. Other Humods that have a cryic temperature regime.

Cryohumods, p. 232

BCE. Other Humods.

Haplohumods, p. 232

Cryohumods

Distinctions between Typic Cryohumods and other subgroups

Typic Cryohumods are the Humods that
a. Have 6 percent or more organic carbon (weighted average) in the matrix of the upper 30 cm of the spodic horizon or, if the spodic horizon is <30 cm thick, in the 30 cm directly below the top of the spodic horizon;
b. Do not have a lithic contact within 50 cm of the mineral soil surface;
c. Do not have an intermittent placic horizon in the spodic horizon;
d. Do not have an argillic or kandic horizon below the spodic horizon;
e. Have a mean annual soil temperature higher than 0°C.
Haplic Cryohumods are like Typic Cryohumods except for *a*.
Lithic Cryohumods are like Typic Cryohumods except for *b*, with or without *a* or *e*, or both.
Pergelic Cryohumods are like Typic Cryohumods except for *e* or for *a* and *e*.

Fragihumods

These are the Humods that have a fragipan below the spodic horizon and do not have a placic horizon. They are not known to occur in the United States, and the classification of subgroups has not been developed.

Haplohumods

Distinctions between Typic Haplohumods and other subgroups

Typic Haplohumods are the Haplohumods that
a. Have either
    (1) A spodic horizon that has a weighted average of 0.6 percent or more organic carbon in the matrix of the upper 30 cm of the spodic horizon or below any Ap horizon; or
    (2) A black upper subhorizon of the spodic horizon that has 3 percent or more organic carbon in the upper 2 cm and that is continuous or is present in >90 percent of the area of each pedon;
b. Have <5 percent by volume of iron-cemented nodules, 2.5 to 30 cm in diameter, in any subhorizon of the spodic horizon;
c. Do not have a lithic contact within 50 cm of the soil surface;
d. Do not have an argillic or kandic horizon below the spodic horizon;
e. Do not have a surface horizon >30 cm thick that meets all the requirements for a plaggen epipedon except thickness;
f. Have a udic moisture regime; and

g. Do not have a sandy epipedon (loamy fine sand or coarser throughout) that is 75 cm or more thick.

Arenic Haplohumods are like Typic Haplohumods except for *g*, and the upper boundary of the spodic horizon is between 75 cm and 1.25 m below the soil surface.

Arenic Ultic Haplohumods are like Typic Haplohumods except for *g* and *d*, with or without *a*, and the upper boundary of the spodic horizon is between 75 cm and 1.25 m below the soil surface.

Entic Haplohumods are like Typic Haplohumods except for *a* or for *a* and *g*, and the upper boundary of the spodic horizon is within 1.25 m of the soil surface.

Ferrudalfic Haplohumods are like Typic Haplohumods except for *b*.

Grossarenic Entic Haplohumods are like Typic Haplohumods except for *a* and *g*, and the upper boundary of the spodic horizon is between 1.25 m and 2 m below the soil surface.

Lithic Haplohumods are like Typic Haplohumods except for *c*.

Orthic Haplohumods are like Typic Haplohumods except for *a(2)*, and the subhorizon is present in <90 percent of the pedon.

Plaggeptic Haplohumods are like Typic Haplohumods except for *e*.

Ultic Haplohumods are like Typic Haplohumods except for *d* or for *a* and *d*.

Xeric Haplohumods are like Typic Haplohumods except for *f*, and they have a xeric moisture regime.

Placohumods

Distinctions between Typic Placohumods and other subgroups

Typic Placohumods are the Placohumods that
a. Have a frigid or warmer temperature regime.
Cryic Placohumods are like the Typic Placohumods in defined properties except for *a*. They have, however, more organic carbon than Typic Placohumods, and in that way they resemble Cryohumods.

Tropohumods

Tropohumods are the Humods that have an isomesic or a warmer iso temperature regime.

Orthods

Key to great groups

BDA. Orthods that have a placic horizon in the spodic horizon.

Placorthods, p. 236

BDB. Other Orthods that have a fragipan below the spodic horizon.

Fragiorthods, p. 234

BDC. Other Orthods that have a cryic or pergelic temperature regime.

Cryorthods, p. 234

BDD. Other Orthods that have an isomesic or warmer iso temperature regime.

Troporthods, p. 236

BDE. Other Orthods.

Haplorthods, p. 235

Cryorthods

Distinctions between Typic Cryorthods and other subgroups

Typic Cryorthods are the Cryorthods that

a. Do not have an argillic or kandic horizon below the spodic horizon;
b. Have a cemented or indurated spodic horizon or have 1.2 to 6 percent organic carbon in the upper 10 cm of the spodic horizon;
c. Do not have a lithic contact within 50 cm of the soil surface;
d. Have a mean annual soil temperature higher than 0°C.
Boralfic Cryorthods are like Typic Cryorthods except for *a* or for *a* and *b*.
Entic Cryorthods are like Typic Cryorthods except for *b* in that they contain less than 1.2 percent organic carbon in the upper 10 cm of the spodic horizon.
Humic Cryorthods are like Typic Cryorthods except for *b* in that they contain >6 percent organic carbon in the upper 10 cm of the spodic horizon.
Humic Lithic Cryorthods are like Typic Cryorthods except for *c* and *b* and they contain >6 percent organic carbon in the upper 10 cm of the spodic horizon.
Lithic Cryorthods are like Typic Cryorthods except for *c* or for *c* and *d*.
Pergelic Cryorthods are like Typic Cryorthods except for *d*.

Fragiorthods

Distinctions between Typic Fragiorthods and other subgroups

Typic Fragiorthods are the Fragiorthods that
a. Do not have an argillic or kandic horizon below the spodic horizon;
b. Do not have distinct or prominent mottles in the spodic horizon;
c. Have a spodic horizon that has one or more of the following:
    (1) A continuous horizon that is at least 2.5 cm thick and is very firm or extremely firm when moist (ortstein);
    (2) A texture of very fine sand or finer and is more than 10 cm thick and has at least 1.2 percent organic carbon (weighted average) in the upper 10 cm;
    (3) A coarse-loamy, loamy-skeletal, sandy-skeletal, or sandy particle-size class and has color value and chroma of 3 or less in at least the upper 7.5 cm;
d. Have a temperature regime warmer than that of Cryorthods;.
e. Do not have an intermittent upper black subhorizon of the spodic horizon that has a ratio of free iron (elemental) to carbon that is <0.2; or, if plowed and the Ap horizon rests

directly on the spodic horizon, do not have tongues of such
a subhorizon;
f. If plowed and the upper part of the spodic horizon thus is
mixed in the Ap horizon and the soil does not have a continuous albic horizon, have more than 1.2 percent organic
carbon in the Ap horizon; and
g. Do not have a surface horizon >30 cm thick that meets all
the requirements for a plaggen epipedon except thickness.
Alfic Fragiorthods are like Typic Fragiorthods except for *a*,
with or without *c* or *f*, or both, and have base saturation of
35 percent or more in some part of the argillic or kandic
horizon or have a mean annual soil temperature less than
8°C.
Aquentic Fragiorthods are like Typic Fragiorthods except for
b and *c* or for *b*, *c*, and *f*.
Aquic Fragiorthods are like Typic Fragiorthods except for *b*.
Cryic Fragiorthods are like Typic Fragiorthods except for *d*.
Entic Fragiorthods are like Typic Fragiorthods except for *c*.
Humic Fragiorthods are like Typic Fragiorthods except for
e.
Plaggeptic Fragiorthods are like Typic Fragiorthods except
for *g*.

Haplorthods

Distinctions between Typic Haplorthods and other subgroups

Typic Haplorthods are the Haplorthods that
a. Do not have an argillic or kandic horizon below the
spodic horizon;
b. Have a spodic horizon that has one or more of the following:
    (1) A continuous horizon at least 2.5 cm thick that is
    very firm or extremely firm when moist (ortstein);
    (2) A texture of very fine sand or finer and is more than
    10 cm thick and has a weighted average of at least 1.2
    percent organic carbon in the upper 10 cm; or
    (3) A coarse-loamy, loamy-skeletal, sandy-skeletal, or
    sandy particle-size class and color value and chroma,
    moist, of 3 or less in at least the upper 7.5 cm of the
    spodic horizon;
c. Do not have distinct or prominent mottles of approximate
spherical shape in the spodic horizon unless the variability
in color is associated with differences in consistence in such
a manner that the redder or darker parts are extremely firm
or very firm, or, if the color is due to uncoated sand grains,
do not have the water table within 1 m of the soil surface
for as many as 60 days, cumulative, in most years;
d. Do not have chroma of 2 or less if mottled or chroma less
than 2 if not mottled, that is dominant in the matrix within
15 cm below the base of the spodic horizon and within 1 m
of the surface of the soil;
e. Do not have a horizon 15 cm or more thick below the
spodic horizon and within 1 m of the surface that has a
brittle matrix when wet or contains some durinodes;
f. Do not have a lithic contact within 50 cm of the surface;
g. Do not have a black intermittent upper subhorizon that
has a ratio of free iron (elemental) to carbon that is <0.2;
h. Have <6 percent organic carbon in the upper 10 cm of
the spodic horizon;

i. Have 1.2 percent or more organic carbon in the Ap horizon if the Ap horizon extends into the upper part of the spodic horizon.

Alfic Haplorthods are like Typic Haplorthods except for *a* or for *a* and *b*, with or without *i*, and the argillic or kandic horizon either has base saturation of 35 percent or more in some part or has a mean annual soil temperature lower than 8°C.

Aqualfic Haplorthods are like Typic Haplorthods except for *a* and *c*, *a* and *d*, or *a*, *c* and *d*, and the argillic or kandic horizon either has base saturation of 35 percent or more in some part or has a mean annual soil temperature lower than 8°C.

Aquentic Haplorthods are like Typic Haplorthods except for *b* and *c* with or without *d*.

Aquic Haplorthods are like Typic Haplorthods except for *c* or *d* or for both *c* and *d*.

Duric Haplorthods are like Typic Haplorthods except for *e* or for *c* and *e*.

Entic Haplorthods are like Typic Haplorthods except for *b*, with or without *i*.

Entic Lithic Haplorthods are like Typic Haplorthods except for *f* and *b*, with or without *i*.

Humic Haplorthods are like Typic Haplorthods except for *g* or *h*.

Lithic Haplorthods are like Typic Haplorthods except for *f*.

Ultic Haplorthods are like Typic Haplorthods except for *a* or for *a* and *b*, with or without *i*, and the argillic or kandic horizon has base saturation throughout of <35 percent and has a mean annual soil temperature of 8°C or more.

Placorthods

These are the orthods that have a placic horizon in the spodic horizon. They are not known to occur in the United States, and they are thought to be rare elsewhere in the world. Subgroups have not been defined.

Troporthods

These are the Orthods that have an isomesic or warmer iso temperature regime. They are not known to occur in the United States, but the group is provided for use elsewhere. Subgroups have not been defined.

Chapter 13

Ultisols

Key To Suborders

FA. Ultisols, either saturated with water at some time of year or artificially drained, that have characteristics associated with wetness, namely, mottles, iron-manganese concretions >2 mm in diameter, or chroma, moist, of 2 or less immediately below any Ap or A horizon that has a value, moist, of less than 3.5 when rubbed; and also one or more of the following:
1. Dominant chroma, moist, of 2 or less in coatings on the surface of peds and mottles within the peds, or dominant chroma of 2 or less in the matrix of the argillic or kandic horizon and mottles of higher chroma (if the hue is redder than 10YR because of parent materials that remain red after citrate-dithionite extraction, the requirement for low chroma is waived);
2. Chroma, moist, of 1 or less on surfaces of peds or in the matrix of the argillic or kandic horizon; or
3. Dominant hue of 2.5Y or 5Y in the matrix of the argillic or kandic horizon and distinct or prominent mottles and also a thermic or isothermic or warmer soil temperature regime.
Aquults, p. 237

FB. Other Ultisols that have either or both of the following characteristics:
1. Have 0.9 percent or more organic carbon in the upper 15 cm of the argillic horizon; or
2. Have 12 kg or more organic carbon in the soil per square meter to a depth of 1 m below the top of the mineral soil surface, exclusive of any O horizon that may be present.
Humults, p. 241

FC. Other Ultisols that have a udic moisture regime.
Udults, p. 245

FD. Other Ultisols that have an ustic moisture regime.
Ustults, p. 252

FE. Other Ultisols that have a xeric moisture regime.
Xerults, p. 257

Aquults

Key to great groups

FAA. Aquults that have plinthite that forms a continuous phase or constitutes more than half the matrix of some subhorizon within 1.5 m of the soil surface.
Plinthaquults, p. 241

FAB. Other Aquults that have a fragipan and, if there is 5 percent or more by volume of plinthite in some subhorizon, the upper boundary of the fragipan is within 1 m of the surface of the soil.
Fragiaquults, p. 239

FAC. Other Aquults that have an abrupt textural change between the ochric epipedon or the albic horizon and the argillic or kandic horizon and have slow hydraulic conductivity in the argillic or kandic horizon.

Albaquults, p. 238

FAD. Other Aquults that

1. Have a CEC $\leq$16 cmol(+) kg^{-1} clay (by 1 M NH$_4$OAc pH 7) and an ECEC $\leq$12 cmol(+) kg^{-1} clay (sum of bases extracted with 1 M NH$_4$OAc pH 7 plus 1 M KCl-extractable Al) in the major part of the argillic or kandic horizon or the major part of the upper 100 cm of the argillic or kandic horizon if these horizons are thicker than 100 cm;
2. Do not have a lithic, paralithic, or petroferric contact within 150 cm of the soil surface; and
3. Have a clay distribution such that the percentage of clay does not decrease from its maximum by as much as 20 percent within a depth of 150 cm from the soil surface or the layer in which the clay percentage decreases by more than 20 percent has at least 5 percent of the volume consisting of skeletans on faces of peds and there is at least 3 percent (absolute) increase in clay below the layer.

Kandiaquults, p. 239

FAE. Other Aquults that have a CEC $\leq$16 cmol(+) kg^{-1} clay (by 1 M NH$_4$OAc pH 7) and an ECEC $\leq$12 cmol(+) kg^{-1} clay (sum of bases extracted with 1 M NH$_4$OAc pH 7 plus 1 M KCl-extractable Al) in the major part of the argillic or kandic horizon or the major part of the upper 100 cm of the argillic or kandic horizon if these horizons are thicker than 100 cm.

Kanhaplaquults, p. 240

FAF. Other Aquults that have a clay distribution such that the percentage of clay does not decrease from its maximum amount by as much as 20 percent within a depth of 150 cm from the soil surface and do not have a lithic or the layer in which the clay percentage decreases by more than 20 percent has at least 5 percent of the volume consisting of skeletans on faces of peds and there is at least 3 percent (absolute) increase in clay below the layer.

Paleaquults, p. 240

FAG. Other Aquults that have an ochric epipedon.

Ochraquults, p. 240

FAH. Other Aquults that have an umbric or a mollic epipedon.

Umbraquults, p. 241

Albaquults

Description of subgroups

Only two subgroups of Albaquults have been proposed, Typic Albaquults and Aeric Albaquults.

Typic Albaquults.--These are the Albaquults that have the colors definitive for the suborder in 60 percent or more of the matrix between the Ap horizon and a depth of 75 cm.

They are moderately extensive in the United States on parts
of the coastal plains in the southeastern states.

Aeric Albaquults.--These soils are not known to occur in
significant areas in the United States, but the subgroup is
provided for use if needed. Either they have the colors that
are definitive for the suborder on ped faces but not in the
ped interiors or they have an ochric epipedon that has
higher chroma or a redder hue, or both.

Fragiaquults

Distinctions between Typic Fragiaquults and other subgroups

Typic Fragiaquults are the Fragiaquults that
a. Have an ochric epipedon;
b. Have mottles and have dominant chroma of 2 or less in
all horizons between the A or Ap horizon and the fragipan;
and
c. Have <5 percent plinthite (by volume) in all subhorizons
within 1.5 m of the soil surface.
Aeric Fragiaquults are like Typic Fragiaquults except for *b*.
Plinthic Fragiaquults are like Typic Fragiaquults except for
c and have lenticular platy structure in the fragipan.
Plinthudic Fragiaquults are like Typic Fragiaquults except
for *b* and *c* and have lenticular platy structure in the fragipan.

Kandiaquults

Distinctions between Typic Kandiaquults and other subgroups

Typic Kandiaquults are the Kandiaquults that
a. Do not have a subhorizon that has dominant chroma of 3
or more within 75 cm of the soil surface;
b. Do not have an epipedon that is 50 cm to 100 cm thick if
the particle-size class is sandy throughout;
c. Do not have an epipedon that is >100 cm thick if the
particle-size class is sandy throughout;
d. Do not have a horizon within 150 cm of the soil surface
that has >5 percent plinthite by volume;
e. Have an ochric epipedon; and
f. Have an ECEC (sum of bases plus 1 *M* KCl-extractable
Al) of more than 1.5 cmol(+) kg^{-1} clay in all subhorizons to
a depth of 150 cm below the soil surface.
Acric Kandiaquults are like Typic Kandiaquults except for *f*,
with or without *a* or *d*, or both.
Aeric Kandiaquults are like Typic Kandiaquults except for *a*.
Arenic Kandiaquults are like Typic Kandiaquults except for
b.
Arenic Plinthic Kandiaquults are like Typic Kandiaquults
except for *b* and *d*.
Arenic Umbric Kandiaquults are like Typic Kandiaquults
except for *b* and *e*.
Grossarenic Kandiaquults are like Typic Kandiaquults except
for *c*.
Plinthic Kandiaquults are like Typic Kandiaquults except for
d.
Umbric Kandiaquults are like Typic Kandiaquults except for
e.

Kanhaplaquults

Distinctions between Typic Kanhaplaquults and other subgroups

Typic Kanhaplaquults are the Kanhaplaquults that
a. Have an ochric epipedon;
b. Do not have a subhorizon that has dominant chroma of 3 or more within 75 cm of the soil surface;
c. Do not have a horizon within 150 cm of the soil surface that has >5 percent plinthite by volume; and
d. Do not have a layer in the upper 75 cm that has texture finer than loamy fine sand, that is as much as 18 cm thick, that has bulk density (at 33 kPa water tension) of 0.95 g per cubic centimeter or less in the fine-earth fraction, and that has either of the following:
  1. A ratio of measured clay to 1500 kPa water (percentages) of 1.25 or less; or
  2. A ratio of CEC (pH 8.2) to 1500 kPa water of <1.5 and more exchange acidity than the sum of bases plus 1 *M* KCl-extractable aluminum.
Aeric Kanhaplaquults are like Typic Kanhaplaquults except for *b*.
Aeric Umbric Kanhaplaquults are like Typic Kanhaplaquults except for *a* and *b*.
Andic Kanhaplaquults are like Typic Kanhaplaquults except for *d*, with or without *a* or *b*, or both.
Plinthic Kanhaplaquults are like Typic Kanhaplaquults except for *c*, with or without *b*.
Umbric Kanhaplaquults are like Typic Kanhaplaquults except for *a*.

Ochraquults

Distinctions between Typic Ochraquults and other Subgroups

Typic Ochraquults are the Ochraquults that:
a. Have dominant chroma of 2 or less in all subhorizons between the A or Ap horizon and a depth of 75 cm.
b. Have texture finer than loamy fine sand in some subhorizon within 50 cm. of the surface.
Aeric Ochraquults are like Typic Ochraquults except for *a*.
Arenic Ochraquults are like Typic Ochraquults except for *b*, with or without *a*, and they have a sandy particle-size class to a depth of between 50 cm and 1m.

Paleaquults

Distinctions between Typic Paleaquults and other subgroups

Typic Paleaquults are the Paleaquults that
a. Do not have a horizon that has dominant chroma of 3 or more within 75 cm of the soil surface;
b. Do not have an epipedon as thick as 50 cm if the particle-size class is sandy throughout;
c. Have <5 percent plinthite (by volume) in all horizons within 1.5 m of the soil surface; and
d. Have an ochric epipedon.
Aeric Paleaquults are like Typic Paleaquults except for *a*.

Arenic Paleaquults are like Typic Paleaquults except for *b*, and they have a sandy particle-size class to a depth between 50 cm and 1 m.
Arenic Plinthic Paleaquults are like Typic Paleaquults except for *b* and *c*, and they have a sandy particle-size class to a depth between 50 cm and 1 m.
Arenic Umbric Paleaquults are like Typic Paleaquults except for *b* and *d*, and they have a sandy particle-size class to a depth between 50 cm and 1 m.
Grossarenic Paleaquults are like Typic Paleaquults except for *b*, and they have a sandy particle-size class to a depth of ≥ 1 m in half or more of the pedon.
Plinthic Paleaquults are like Typic Paleaquults except for *c*.
Umbric Paleaquults are like Typic Paleaquults except for *d*.

Plinthaquults

Distinction between Typic Plinthaquults and other subgroups.

Typic Plinthaquults are the Plinthaquults that
a. Have a CEC of 24 or more cmol(+) kg^{-1} of clay (by 1 *M* NH$_4$OAc pH7) in the major part of the argillic or kandic horizon or the major part of the upper 100 cm of the argillic or kandic horizon if these horizons are thicker than 100 cm.
Kandic Plinthaquults are like Typic Plinthaquults except for *a.*

Umbraquults

Umbraquults are the Aquults that
1. Have an umbric or mollic epipedon;
2. Do not have a fragipan;
3. Do not have plinthite that forms a continuous phase or constitutes more than half of the matrix of any subhorizon within 150 cm of the soil surface;
4. Have a clay distribution such that the percentage of clay decreases from its maximum by 20 percent or more of the maximum within a depth of 150 cm below the soil surface, and the layer in which the percentage of clay is less than the maximum has less than 5 percent of the volume consisting of skeletans on faces of peds or there is less than 3 percent (absolute) increase in clay below the layer.

Humults

Key to great groups

FBA. Humults that have a sombric horizon within 1 m of the soil surface.

Sombrihumults, p. 244

FBB. Other Humults that have plinthite that forms a continuous phase or constitutes >50 percent of the volume of some subhorizon within 150 cm of the soil surface.

Plinthohumults, p. 244

FBC. Other Humults that
1. Have a CEC ≤ 16 cmol(+) kg^{-1} clay (by 1 *M* NH$_4$OAc pH 7) and an ECEC ≤ 12 cmol(+) kg^{-1} clay (sum of bases ex-

tracted with 1 M NH$_4$OAc pH 7 plus 1 M KCl-extractable
Al) in the major part of the argillic or kandic horizon or the
major part of the upper 100 cm of the argillic or kandic
horizon if these horizons are thicker than 100 cm;
2. Do not have a lithic, paralithic, or petroferric contact
within 150 cm of the soil surface; and
3. Have a clay distribution such that the percentage of clay
does not decrease from its maximum amount by as much as
20 percent within a depth of 150 cm from the soil surface
or the layer in which the clay percentage decreases by more
than 20 percent has at least 5 percent of the volume consisting of skeletans on faces of peds and there is at least 3
percent (absolute) increase in clay below the layer.

Kandihumults, p. 243

FBD. Other Humults that have a CEC $\leq$16 cmol(+) kg^{-1} clay
(by 1 M NH$_4$OAc pH 7) and an ECEC $\leq$12 cmol(+) kg^{-1} clay
(sum of bases extracted with 1 M NH$_4$OAc pH 7 plus 1 M
KCl-extractable Al) in the major part of the argillic or
kandic horizon or the major part of the upper 100 cm of the
argillic or kandic horizon if these horizons are thicker than
100 cm.

Kanhaplohumults, p. 243

FBE. Other Humults.

Haplohumults, p. 242

Haplohumults

Distinctions between Typic Haplohumults and other subgroups

Typic Haplohumults are the Haplohumults that
a. Do not have a layer in the upper 75 cm that has texture
finer than loamy fine sand, that is as much as 18 cm thick,
that has a bulk density (at 33 kPa water tension) of 0.95 g
or less per cubic centimeter in the fine-earth fraction, and
that has either of the following:
    (1) A ratio of measured clay to 1500 kPa water
    (percentages) of 1.25 or less; or
    (2) A ratio of CEC (at pH near 8) to 1500 kPa water of
    >1.5 and more exchange acidity than the sum of bases
    plus KCl-extractable aluminum;
b. Have a udic moisture regime;
c. Do not have the following combination of characteristics
in the upper 25 cm or more of the argillic horizon:
    (1) Mottles that have a color value, moist, of 4 or more
    and chroma, moist, of 2 or less accompanied by mottles
    of higher chroma that are due to segregation of iron; and
    (2) Saturation with water in the mottled zone at some
    time of year when the soil temperature in that zone is
    5°C or more or artificial drainage; and
d. Do not have a lithic contact within 50 cm of the mineral
soil surface.
Andeptic Haplohumults are like Typic Haplohumults except
for *a.*
Aquic Haplohumults are like Typic Haplohumults except for
c.
Xeric Haplohumults are like Typic Haplohumults except for
b and have a xeric soil moisture regime.

Kandihumults

Distinctions between Typic Kandihumults and other subgroups

Typic Kandihumults are the Kandihumults that
a. Do not have the following combination of characteristics
in the upper 25 cm or more of the argillic or kandic horizon:
    1. Mottles that have a color value, moist, of 4 or more,
    and chroma, moist, of 2 or less accompanied by mottles
    of higher chroma that are due to segregation of iron; and
    2. Saturation with water in the mottled zone at some tim
    of year when the soil temperature in that zone is 5°C or
    more or there is artificial drainage;
b. Have an udic moisture regime;
c. Do not have a layer in the upper 75 cm that has texture
finer than loamy fine sand, that is as much as 18 cm thick,
that has a bulk density (at 33 kPa water tension) of 0.95 g
per cubic centimeter or less in the fine-earth fraction, and
that has either of the following:
    1. A ratio of measured clay to 1500 kPa water
    (percentages) of 1.25 or less; or
    2. A ratio of CEC (pH 8.2) to 1500 kPa water of >1.5
    and more exchange acidity than the sum of bases plus
    KCl-extractable aluminum;
d. Do not have a horizon within 150 cm of the soil surface
that has >5 percent plinthite by volume;
e. Have a hue redder than 10YR in all parts of the soil
above a depth of 75 cm that have a color value, moist, of 4
or more if there are mottles of high chroma within that
depth and if the hue becomes redder with depth within 100
cm of the soil surface; and
f. Do not have an anthropic epipedon.
Andic Kandihumults are like Typic Kandihumults except for
c.
Andic Epiaquic Kandihumults are like Typic Kandihumults
except for *c* and *e.*
Andic Ustic Kandihumults are like Typic Kandihumults
except for *b* and *c* and they have an ustic moisture regime.
Anthropic Kandihumults are like Typic Kandihumults except
for *f.*
Epiaquic Kandihumults are like Typic Kandihumults except
for *e.*
Plinthic Kandihumults are like Typic Kandihumults except
for *d.*
Ustic Kandihumults are like Typic Kandihumults except for
b and they have an ustic moisture regime.

Kanhaplohumults

Distinctions between Typic Kanhaplohumults and other subgroups

Typic Kanhaplohumults are the Kanhaplohumults that
a. Do not have the following combination of characteristics
in the upper 25 cm or more of the argillic or kandic horizon;
    1. Mottles that have a color value, moist, of 4 or more
    and chroma, moist, of 2 or less accompanied by mottles
    of higher chroma that are due to segregation of iron; and

2. Saturation with water in the mottled zone at some time of year when the soil temperature in that zone is 5°C or higher or there is artificial drainage;

b. Do not have a lithic contact within 50 cm or the mineral soil surface;

c. Have an udic moisture regime;

d. Do not have a layer in the upper 75 cm that has texture finer than loamy fine sand, that is as much as 18 cm thick, that has bulk density (at 33 kPa water tension) of 0.95 g per cubic centimeter or less in the fine-earth fraction, and that has either of the following:

1. A ratio of measured clay to 1500 kPa water (percentages) of 1.25 or less; or

2. A ratio of CEC (pH 8.2) to 1500 kPa water of >1.5 and more exchange acidity than the sum of bases plus KCl-extractable aluminum;

e. Have a hue redder than 10YR in all parts of the soil above a depth of 75 cm that have a color value, moist, of 4 or more if there are mottles of high chroma above that depth and if the hue becomes redder with depth within 100 cm of the soil surface; and

f. Do not have an anthropic epipedon.

Andic Kanhaplohumults are like Typic Kanhaplohumults except for *d*, with or without *e* or *f*.

Andic Ustic Kanhaplohumults are like Typic Kanhaplohumults except for *c* and *d*, with or without *e* or *f*, and they have an ustic moisture regime.

Anthropic Kanhaplohumults are like Typic Kanhaplohumults except for *f*.

Aquic Kanhaplohumults are like Typic Kanhaplohumults except for *a* or for *a* and *e*.

Epiaquic Kanhaplohumults are like Typic Kanhaplohumults except for *e*.

Lithic Kanhaplohumults are like Typic Kanhaplohumults except for *b* with or without any of *a*, *c*, *d*, *e*, or *f*.

Ustic Kanhaplohumults are like Typic Kanhaplohumults except for *c* and they have an ustic moisture regime.

Plinthohumults

These are the Humults that have plinthite that forms a continuous phase or that constitutes more than half the volume of some subhorizon within 1.25 m of the soil surface. They are not known to occur in the United States, but the great group has been proposed for other countries (Sys 1969). Subgroups have not been developed.

Sombrihumults

These are the Humults that have a sombric horizon whose upper boundary is within 1 m of the soil surface. They are not known to occur in the United States, but the great group is provided for use elsewhere. Humoxic and orthoxic subgroups have been proposed (Sys 1969) for the soils that have low CEC, that is, <24 cmol(+) kg^{-1} clay. The humoxic subgroup is proposed for soils that have an isothermic or cooler temperature regime and the orthoxic subgroup for soils that have an isohyperthermic temperature regime.

Udults

Key to great groups

FCA. Udults that have plinthite that forms a continuous
phase or constitutes more than half the volume in some subhorizon within 150 cm of the soil surface.

Plinthudults, p. 251

FCB. Other Udults that have a fragipan in or below the
argillic or kandic horizon.

Fragiudults, p. 246

FCC. Other Udults that

1. Have a CEC $\leq$16 cmol(+) kg^{-1} clay (by 1 M NH$_4$OAc pH
7) and an ECEC $\leq$12 cmol(+) kg^{-1} clay (sum of bases extracted with 1 M NH$_4$OAc pH 7 plus 1 M KCl-extractable
Al) in the major part of the argillic or kandic horizon or the
major part of the upper 100 cm of the argillic or kandic
horizon if these horizons are thicker than 100 cm;
2. Do not have a lithic, paralithic, or petroferric contact
within 150 cm of the soil surface: and
3. Have a clay distribution such that the percentage of clay
does not decrease from its maximum amount by as much as
20 percent within a depth of 150 cm from the soil surface
or the layer in which the clay percentage decreases by more
than 20 percent has at least 5 percent of the volume consisting of skeletans on faces of peds and there is at least 3
percent (absolute) increase in clay below the layer.

Kandiudults, p. 247

FCD. Other Udults that have a CEC $\leq$16 cmol(+) kg^{-1} clay
(by 1 M NH$_4$OAc pH 7) and an ECEC $\leq$12 cmol(+) kg^{-1} clay
(sum of bases extracted with 1 M NH$_4$OAc pH 7 plus 1 M
KCl-extractable Al) in the major part of the argillic or
kandic horizon or the major part of the upper 100 cm of the
argillic or kandic horizon if these horizons are thicker than
100 cm.

Kanhapludults, p. 249

FCE. Other Udults that have a clay distribution such that
the percentage of clay does not decrease from its maximum
amount by more than 20 percent within 150 cm of the soil
surface or the layer in which the clay percentage decreases
by more than 20 percent has at least 5 percent of the volume consisting of skeletans on faces of peds and there is at
least 3 percent (absolute) increase in clay below the layer.

Paleudults, p. 250

FCF. Other Udults that have
1. An epipedon that has a color value, moist, less than 3.5 in
all parts; and
2. An argillic horizon that has a color value, dry, less than 5
and not more than 1 unit higher than the value, moist.

Rhodudults, p. 251

FCG. Other Udults.

Hapludults, p. 246

Fragiudults

Distinctions between Typic Fragiudults and other subgroups

Typic Fragiudults are the Fragiudults that
a. Meet these two requirements:
  (1) Have an argillic or kandic horizon above the fragipan that has some clay skins on both vertical and horizontal surfaces of some structural aggregates; and
  (2) Do not have an intervening horizon (one or more) between the argillic or kandic horizon and the fragipan that has dominant chroma of 3 or less and that has as much as 3 percent less clay (absolute) than both the overlying argillic or kandic horizon and the underlying fragipan;
b. Do not have mottles that have chroma of 2 or less:
  (1) Within the upper 25 cm of the argillic or kandic horizon or above the fragipan if the part of the argillic or kandic horizon above the fragipan is less than 25 cm thick; or
  (2) Within 40 cm of the surface of the soil if the argillic or kandic horizon is lacking above the fragipan or there is an intervening horizon between the argillic or kandic horizon and the fragipan that has as much as 3 percent less clay (absolute) than both the overlying argillic or kandic horizon and the underlying fragipan.
c. Do not have an epipedon as thick as 50 cm if the particle-size class is sandy throughout;
d. Have <5 percent plinthite by volume in all horizons within 1.5 m of the soil surface; and
e. Have an Ap horizon that has a color value, moist, of 4 or more or has a value, dry, of 6 or more when crushed and smoothed (smoothed with a knife to eliminate shadows), or the A horizon is <15 cm thick if its color value, moist, is lower than 3.5.
Aquic Fragiudults are like Typic Fragiudults except for *b(1)*.
Arenic Fragiudults are like Typic Fragiudults except for *c* or for *a* and *c*, and they have a sandy epipedon that is 50 cm to 1 m thick.
Glossaquic Fragiudults are like Typic Fragiudults except for *a* and *b(2)*.
Glossic Fragiudults are like Typic Fragiudults except for *a*.
Plinthaquic Fragiudults are like Typic Fragiudults except for *a*, *b(2)*, and *d*.
Plinthic Fragiudults are like Typic Fragiudults except for *a* and *d*.

Hapludults

Distinctions between Typic Hapludults and other subgroups

Typic Hapludults are the Hapludults that
a. Do not have the following combination of characteristics in the upper 60 cm of the argillic horizon:
  (1) Mottles that have a color value, moist, of 4 or more and chroma, moist, of 2 or less, and also mottles of higher chroma that are due to segregation of iron; and
  (2) Saturation with water in the mottled zone at some time of year when the soil temperature in that zone is 5°C or higher, or artificial drainage;

b. Do not have an epipedon as thick as 50 cm if the particle-size class is sandy throughout;
c. Have an argillic horizon >25 cm thick;
d. Have an Ap horizon that has a color value, moist, of 4 or more or has a value, dry, of 6 or more when crushed and smoothed; or the A   horizon is <15 cm thick if its color value, moist, is less than 3.5;
e. Do not have a lithic contact within 50 cm of the surface of the mineral soil;
f. Have texture finer than loamy fine sand in some part of the argillic horizon and have an argillic horizon that, in at least its upper 25 cm, does not have lamellae;
g. Do not have the following combination of characteristics:
   (1) Cracks at some period in most years that are 1 cm or more wide at a depth of 50 cm, that are at least 30 cm long in some part, and that extend upward to the soil surface, to the base of an Ap horizon, or to a depth within 25 cm of the soil surface; and
   (2) A coefficient of linear extensibility (COLE) of 0.09 or more in a horizon or horizons at least 50 cm thick and a potential linear extensibility of 6 cm or more in the upper 1 m of the soil or in the whole soil if a lithic or paralithic contact is deeper than 50 cm but shallower than 1 m; and
   (3) More than 35 percent clay in horizons that total >50 cm in thickness; and
h. Have a continuous argillic horizon throughout each pedon, not interrupted by ledges of bedrock.
Aquic Hapludults are like Typic Hapludults except for *a* or for *a* and *d*.
Arenic Hapludults are like Typic Hapludults except for *b*, with or without *a* or *c* or both, and the epipedon is between 50 cm and 1 m thick.
Humic Hapludults are like Typic Hapludults except for *d*.
Lithic Hapludults are like Typic Hapludults except for *e* or for *c* and *e*.
Ochreptic Hapludults are like Typic Hapludults except for *c*.
Psammentic Hapludults are like Typic Hapludults except for *f* or for *f* and *b*.
Ruptic-Lithic-Entic Hapludults are like Typic Hapludults except for *h* and *e* or for *h*, *c*, and *e*.
Vertic Hapludults are like Typic Hapludults except for *g*, with or without *a*.

Kandiudults

Distinctions between Typic Kandiudults and other subgroups

Typic Kandiudults are the Kandiudults that
a. Do not have the following combination of characteristics in the upper 75 cm of the soil if the chroma throughout the upper 75 cm is not controlled by uncoated sand grains; or if the chroma throughout the upper 75 cm is controlled by uncoated sand grains, do not have the following combination of characteristics throughout the upper 12.5 cm of the argillic or kandic horizon:
   1. Mottles that have a color value, moist, of 4 or more and chroma, moist, of 2 or less accompanied by mottles of higher chroma that are due to segregation of iron; and

2. Saturation with water in the mottled zone at some time of year when the soil temperature in that zone is 5°C or higher, or there is artificial drainage;

b. Do not have an epipedon that is 50 cm to 100 cm thick if the particle-size class is sandy throughout;

c. Do not have an epipedon that is >100 cm thick if the particle-size class is sandy throughout;

d. Do not have a subhorizon within 150 cm of the soil surface that has >5 percent plinthite by volume;

e. Have an argillic or kandic horizon that has a color hue of 5YR or yellower in some part, or has a value, moist, of 3.5 or more or has a value, dry, that is more than 1 unit higher than the value, moist;

f. Do not have a layer in the upper 75 cm that has texture finer than loamy fine sand, that is as much as 18 cm thick, that has bulk density (at 33 kPa water tension) of 0.95 g per cubic centimeter or less in the fine-earth fraction, and that has either of the following:

1. A ratio of measured clay to 1500 kPa water (percentages) of 1.25 or less; or

2. A ratio of CEC (pH 8.2) to 1500 kPa water of >1.5 and more exchange acidity than the sum of bases plus KCl-extractable aluminum;

g. Have a hue redder than 10YR in all parts of the soil above a depth of 75 cm that have color value, moist, of 4 or more if there are mottles of high chroma within that depth and if the hue becomes redder with depth within 100 cm of the soil surface;

h. Have an ECEC (sum of bases plus 1 M KCl-extractable Al) of more than 1.5 cmol(+) kg^{-1} clay in all subhorizons to a depth of 150 cm below the soil surface; and

i. Do not have a sombric horizon within 150 cm of the soil surface.

Acric Kandiudults are like Typic Kandiudults except for *h*, with or without *f*.

Acric Plinthic Kandiudults are like Typic Kandiudults except for *d* and *h*, with or without *f*.

Andaquic Kandiudults are like Typic Kandiudults except for *a* and *f*.

Andic Kandiudults are like Typic Kandiudults except for *f*.

Aquic Kandiudults are like Typic Kandiudults except for *a*.

Aquic Arenic Kandiudults are like Typic Kandiudults except for *a* and *b*.

Arenic Kandiudults are like Typic Kandiudults except for *b*.

Arenic Plinthaquic Kandiudults are like Typic Kandiudults except for *a*, *b*, and *d*.

Arenic Plinthic Kandiudults are like Typic Kandiudults except for *b* and *d*.

Arenic Rhodic Kandiudults are like Typic Kandiudults except for *b* and *e*.

Epiaquic Kandiudults are like Typic Kandiudults except for *g*.

Grossarenic Kandiudults are like Typic Kandiudults except for *c* or for *a* and *c*.

Grossarenic Plinthic Kandiudults are like Typic Kandiudults except for *c* and *d*.

Plinthaquic Kandiudults are like Typic Kandiudults except for *a* and *d*.

Plinthic Kandiudults are like Typic Kandiudults except for *d*.

Rhodic Kandiudults are like Typic Kandiudults except for *e*.

Sombric Kandiudults are like Typic Kandiudults except for
i.

Kanhapludults

Distinctions between Typic Kanhapludults and other subgroups

Typic Kanhapludults are the Kanhapludults that
a. Do not have the following combination of characteristics
in the upper 75 cm of the soil if the chroma throughout the
upper 75 cm is not controlled by the uncoated sand grains;
or if the chroma throughout the upper 75 cm is controlled
by uncoated sand grains, do not have the following combination of characteristics throughout the upper 12.5 cm of
the argillic or kandic horizon:
> 1. Mottles that have a color value, moist, of 4 or more
> and chroma, moist, of 2 or less accompanied by mottles
> of higher chroma that are due to segregation or iron; and
> 2. Saturation with water in the mottled zone at some time
> of year when the soil temperature in that zone is 5°C or
> higher or there is artificial drainage;

b. Do not have an epipedon that is $\geq$50 cm thick if the particle-size class is sandy throughout;
c. Do not have a horizon within 150 cm of the soil surface
that has >5 percent plinthite by volume;
d. Do not have a lithic contact within 50 cm of the soil
surface;
e. Have a hue redder than 10YR in all parts of the soil
above a depth of 75 cm that have color value, moist, of 4 or
more if there are mottles of high chroma within that depth
and if the hue becomes redder with depth within 100 cm of
the soil surface;
f. Have an argillic or kandic horizon that has a color hue of
5YR or yellower in some part, or has a value, moist, of 3.5
or more in some part, or has a value, dry, that is more than
one unit higher than the value, moist;
g. Do not have a layer in the upper 75 cm that has texture
finer than loamy fine sand, that is as much as 18 cm thick,
that has bulk density (at 33 kPa water tension) of 0.95 g per
cubic centimeter or less in the fine-earth fraction and that
has either of the following:
> 1. A ratio of measured clay to 1500 kPa water
> (percentages) of 1.25 or less; or
> 2. A ratio of CEC (pH 8.2) to 1500 kPa water of >1.5
> and more exchange acidity than the sum of bases plus
> KCl-extractable aluminum; and

h. Have an ECEC (sum of bases plus 1 *M* KCl-extractable
Al) of more than 1.5 cmol(+) kg^{-1} clay in all subhorizons to
a depth of 150 cm below the soil surface.
Acric Kanhapludults are like Typic Kanhapludults except for
h, with or without any of *c*, *e*, or *g*.
Andic Kanhapludults are like Typic Kanhapludults except
for *g*, with or without *e*.
Aquic Kanhapludults are like Typic Kanhapludults except for
a.
Arenic Kanhapludults are like Typic Kanhapludults except
for *b*.
Arenic Plinthic Kanhapludults are like Typic Kanhapludults
except for *b* and *c*.

Epiaquic Kanhapludults are like Typic Kanhapludults except for *e*.

Lithic Kanhapludults are like Typic Kanhapludults except for *d*, with or without *a*, *e*, *f*, or *g*.

Plinthaquic Kanhapludults are like Typic Kanhapludults except for *a* and *c*, with or without *g*.

Plinthic Kanhapludults are like Typic Kanhapludults except for *c*, with or without *g*.

Rhodic Kanhapludults are like Typic Kanhapludults except for *f*.

Paleudults

Distinctions between Typic Paleudults and other subgroups

Typic Paleudults are the Paleudults that

a. Do not have the following combination of characteristics in the upper 75 cm of the soil if the chroma throughout the upper 75 cm is not controlled by the uncoated sand grains; or if the chroma throughout the upper 75 cm is controlled by uncoated sand grains, do not have the following combination of characteristics throughout the upper 12.5 cm of the argillic horizon:

    (1) Mottles that have a color value, moist, of 4 or more and chroma, moist, of 2 or less and mottles of higher chroma that are due to segregation of iron; and

    (2) Saturation with water in the mottled zone at some time of year when the soil temperature in that zone is 5°C higher, or artificial drainage;

b. Do not have an epipedon as thick as 50 cm if the particle-size class is sandy throughout;

c. Do not have a subhorizon within 1.5 m of the soil surface that has >5 percent plinthite;

d. Have an argillic horizon that has a color value, moist, of 4 or more or that has mottles of high chroma in some subhorizon within 1 m of the top of the argillic horizon, or have a color value, dry, more than 1 unit higher than the value, moist, in some part of the soil within that depth;

e. Have texture finer than loamy fine sand in some part of the argillic horizon and do not have lamellae in at least the upper 1 m of the argillic horizon;

f. Do not have a horizon that is above the argillic horizon whose lower boundary is deeper than 18 cm and that meets all requirements for a spodic horizon except the horizon is intermittent; and

g. Do not have a subhorizon in the argillic horizon and within 1.25 m of the soil surface that has all the properties of a fragipan except that it is brittle in 40 to 60 percent of the volume.

Aquic Paleudults are like Typic Paleudults except for *a*.

Aquic Arenic Paleudults are like Typic Paleudults except for *a* and
b, and they have a sandy epipedon that is 50 cm to 1 m thick.

Arenic Paleudults are like Typic Paleudults except for *b*, and they have a sandy epipedon that is 50 cm to 1 m thick.

Arenic Plinthaquic Paleudults are like Typic Paleudults except for *a*, *b*, and *c*, with or without *g*. They have a sandy epipedon that is 50 cm to 1 m thick, and have mottles that have chroma of 2 or less and also have high-chroma mottles in the sandy epipedon and in the upper 12.5 cm of the argillic horizon.

Arenic Plinthic Paleudults are like Typic Paleudults except for *b* and *c* or for *b*, *c*, and *g*, and they have a sandy epipedon that is 50 cm to 1 m thick.

Arenic Rhodic Paleudults are like Typic Paleudults except for *b* and *d*, and they have a sandy epipedon that is 50 cm to 1 m thick.

Fragiaquic Paleudults are like Typic Paleudults except for *a* and *g*.

Fragic Paleudults are like Typic Paleudults except for *g*.

Grossarenic Paleudults are like Typic Paleudults except for *b* or for *a* and *b*, and they have a sandy epipedon that is between 1 and 2 m thick in half or more of each pedon.

Grossarenic Plinthic Paleudults are like Typic Paleudults except for *b* and *c* or for *b*, *c*, and *g*, and they have a sandy epipedon that is between 1 and 2 m thick in half or more of each pedon.

Plinthaquic Paleudults are like Typic Paleudults except for *a* and *c*, with or without g.

Plinthic Paleudults are like Typic Paleudults except for *c*, with or without g.

Psammaquentic Paleudults are like Typic Paleudults except for *a* and *e*, with or without *b*.

Psammentic Paleudults are like Typic Paleudults except for *e*.

Rhodic Paleudults are like Typic Paleudults except for *d*.

Spodic Paleudults are like Typic Paleudults except for *f*, with or without *a* or *b*, or both.

Plinthudults

Plinthudults are the Udults that have plinthite that forms a continuous phase or constitutes more than half the matrix of some subhorizon in the upper 150 cm of the soil.

Rhodudults

Distinctions between Typic Rhodudults and other subgroups

Typic Rhodudults are the Rhodudults that
a. Have an argillic horizon that is continuous vertically and horizontally and has a hue redder than 5YR;
b. Have texture finer than loamy fine sand in some part of the argillic horizon; and
c. Do not have a lithic contact within 50 cm of the soil surface.

Ustults

Key to great groups

FDA. Ustults that have plinthite that forms a continuous phase or that constitutes more than half the volume of some subhorizon within 150 cm of the soil surface.
Plinthustults, p. 256

FDB. Other Ustults that
1. Have a CEC $\leq$16 cmol(+) kg^{-1} clay (by 1 M NH$_4$OAc pH 7) and an ECEC $\leq$12 cmol(+) kg^{-1} clay (sum of bases extracted with 1 M NH$_4$OAc pH 7 plus 1 M KCl-extractable

A1) in the major part of the argillic or kandic horizon or the major part of the upper 100 cm of the argillic or kandic horizon if these horizons are thicker than 100 cm;
2. Do not have a lithic, paralithic, or petroferric contact within 150 cm of the soil surface; and
3. Have a clay distribution such that the percentage of clay does not decrease from its maximum amount by as much as 20 percent within a depth of 150 cm from the soil surface or the layer in which the clay percentage decreases by more than 20 percent has at least 5 percent of the volume consisting of skeletans on faces of peds and there is at least 3 percent (absolute) increase in clay below the layer.

Kandiustults, p. 253

FDC. Other Ustults that have a CEC $\leq$16 cmol(+) kg^{-1} clay (by 1 M NH$_4$OAc pH 7) and an ECEC $\leq$12 cmol(+) kg^{-1} clay (sum of bases extracted with 1 M NH$_4$OAc pH 7 plus 1 M KCl-extractable Al) in the major part of the argillic or kandic horizon or the major part of the upper 100 cm of the argillic or kandic horizon if these horizons are thicker than 100 cm.

Kanhaplustults, p. 254

FDD. Other Ustults that have a clay distribution such that the percentage of clay does not decrease from its maximum amount by as much as 20 percent of that maximum within 150 cm of the soil surface or the layer in which the clay percentage decreases by more than 20 percent has at least 5 percent of the volume consisting of skeletans on faces of peds and there is at least 3 percent (absolute) increase in clay below the layer.

Paleustults, p. 256

FDE. Other Ustults that have:
1. An epipedon that has a color value, moist, less than 3.5 in all parts; and
2. An argillic horizon that has a color value, dry, less than 5 and not more than 1 unit higher than the value, moist.

Rhodustults, p. 256

FDF. Other Ustults.

Haplustults, p. 252

Haplustults

Distinctions between Typic Haplustults and other subgroups

Typic Haplustults are the Haplustults that
a. Have a continuous argillic horizon throughout each pedon, uninterrupted by ledges of bedrock;
b. Do not have a lithic contact within 50 cm of the mineral soil surface;
c. Have texture finer than loamy fine sand in some part of the argillic horizon and have an argillic horizon that does not have lamellae in at least its upper 25 cm;
d. Do not have the following combination of characteristics in the upper 75 cm of the soil and in the upper 12.5 cm of the argillic horizon:
    (1) Mottles that have a color value, moist, of 4 or more and chroma, moist, of 2 or less and mottles of higher chroma that are due to segregation of iron; and

(2) Saturation with water in the mottled zone at some
time of year when the soil temperature in that zone is
5°C or higher or artificial drainage;
e. Have a hue redder than 10YR in all parts of the upper 75
cm of soil that have a color value, moist, of 4 or more if
there are mottles of high chroma within that depth and if
the hue becomes redder with depth within 1 m of the soil
surface;
f. Do not have a petroferric contact within 1 m of the soil
surface;
g. Do not have a horizon within 1.5 m of the soil surface
that has $\geq$ 5 percent plinthite by volume;
h. Have CEC of 24 or more cmol(+) kg^{-1} clay (by 1 M
NH$_4$OAc pH7) in the major part of the argillic horizon or
the major part of the upper 100 cm of the argillic horizon if
the argillic horizon is thicker than 100 cm; and
i. Do not have an epipedon as thick as 50 cm if its particle-
size class is sandy throughout.
Aquic Haplustults are like Typic Haplustults except for *d.*
Arenic Haplustults are like Typic Haplustults except for *i,*
and they have a sandy epipedon that is 50 cm to 1 m thick.
Epiaquic Haplustults are like Typic Haplustults except for *e.*
Kanhaplic Haplustults are like Typic Haplustults except for
h.
Lithic Haplustults are like Typic Haplustults except for *b.*
Petroferric Haplustults are like Typic Haplustults except for
f.
Plinthic Haplustults are like Typic Haplustults except for *g.*

Kandiustults

Distinctions between Typic Kandiustults and other subgroups

Typic Kandiustults are the Kandiustults that
a. Do not have the following combination of characteristics
in the upper 75 cm of the soil if the chroma throughout the
upper 75 cm is not controlled by uncoated sand grains; or if
the chroma throughout the upper 75 cm is controlled by
uncoated sand grains, do not have the following combination
of characteristics throughout the upper 12.5 cm of the
argillic or kandic horizon:
  1. Mottles that have a color value, moist, of 4 or more
  and chroma, moist, of 2 or less accompanied by mottles
  of higher chroma that are due to segregation of iron; and
  2. Saturation with water in the mottled zone at some time
  of year when the soil temperature in that zone is 5°C or
  higher or there is artificial drainage;
b. Do not have a horizon within 150 cm of the soil surface
that has >5 percent plinthite by volume;
c. Do not have an epipedon that is $\geq$50 cm thick if the particle-size class is sandy throughout;
d. Have an argillic or kandic horizon that has a color hue of
5YR or yellower in some part, or has a value, moist, of 3.5
or more in some part, or has a value, dry, that is more than
one unit higher than the value, moist;
e. Do not have a layer in the upper 75 cm that has texture
finer than loamy fine sand, that is as much as 18 cm thick,
that has a bulk density (at 33 kPa water tension) of 0.95 g
per cubic centimeter or less in the fine-earth fraction, and
that has either of the following:

1. A ratio of measured clay to 1500 kPa water (percentages) of 1.25 or less; or
2. A ratio of CEC (pH 8.2) to 15-water of >1.5 and more exchange acidity than the sum of bases plus KCl-extractable aluminum;
f. When neither irrigated nor fallowed to store moisture:
1. If the soil temperature regime is mesic or thermic, are moist more than six-tenths of the time in half or more years in some part of the moisture control section (not necessarily the same part) when the soil temperature at a depth of 50 cm exceeds 5°C; or
2. If the soil temperature regime is hyperthermic, isomesic, or warmer, are moist in most years in some or all parts of the moisture control section for 180 or more days during a period when the soil temperature at a depth of 50 cm exceeds 8°C; and
g. When neither irrigated nor fallowed to store moisture:
1. If the soil temperature regime is mesic or thermic, are dry for more than 135 cumulative days in some part of the moisture control section when the soil temperature at a depth of 50 cm exceeds 5°C; or
2. If the soil temperature regime is hyperthermic, isomesic, or warmer, the soils are dry in some or all parts of the moisture control section for more than 90 days during a period when the soil temperature at a depth of 50 cm exceeds 8°C; and
h. Have an ECEC (sum of bases plus KCl-extractable Al) of more than 1.5 cmol(+) kg^{-1} clay in all subhorizons to a depth of 150 cm below the soil surface.
Acric Kandiustults are like Typic Kandiustults except for *h* with or without *a*, *b*, or *d*.
Andic Kandiustults are like Typic Kandiustults except for *e*, with or without *b*.
Andic Udic Kandiustults are like Typic Kandiustults except for *e* and *g*.
Aquic Kandiustults are like Typic Kandiustults except for *a*.
Arenic Kandiustults are like Typic Kandiustults except for *c*.
Arenic Plinthic Kandiustults are like Typic Kandiustults except for *b* and *c*.
Aridic Kandiustults are like Typic Kandiustults except for *f*.
Plinthic Kandiustults are like Typic Kandiustults except for *b*.
Rhodic Kandiustults are like Typic Kandiustults except for *d*.
Udic Kandiustults are like Typic Kandiustults except for *g*.

Kanhaplustults

Distinctions between Typic Kanhaplustults and other subgroups

Typic Kanhaplustults are the Kanhaplustults that
a. Do not have a lithic contact within 50 cm of the mineral soil surface;
b. Do not have the following combination of characteristics in the upper 75 cm of the soil if the chroma throughout the upper 75 cm is not controlled by uncoated sand grains; or if the chroma throughout the upper 75 cm is controlled by uncoated sand grains, do not have the following combination of characteristics throughout the upper 12.5 cm of the argillic or kandic horizon:

1. Mottles that have a color value, moist, of 4 or more
and chroma. moist, of 2 or less accompanied by mottles
of higher chroma that are due to segregation of iron; and
2. Saturation with water in the mottled zone at some time
of year when the soil temperature in that zone is 5°C or
higher or there is artificial drainage;
c. Have a hue redder than 10YR in all parts of the soil
above a depth of 75 cm that have a color value, moist, of 4
or more if there are mottles of high chroma above that
depth and if the hue becomes redder with depth within 100
cm of the soil surface;
d. Do not have a horizon within 150 cm of the soil surface
that has >5 percent plinthite by volume;
e. Do not have an epipedon that is ≥50 cm thick if the particle-size class is sandy throughout;
f. Have an argillic or kandic horizon that has a color hue of
5YR or yellower in some part, or has a value, moist, of 3.5
or more in some part, or has a value, dry, that is more than
one unit higher than the value, moist;
g. Do not have a layer in the upper 75 cm that has texture
finer than loamy fine sand, that is as much as 18 cm thick,
that has bulk density (at 33 kPa water tension) of 0.95 g per
cubic centimeter or less in the fine-earth fraction, and that
has either of the following:
1. A ratio of measured clay to 1500 kPa water
(percentages) of 1.25 or less; or
2. A ratio of CEC (pH 8.2) to 1500 kPa water of >1.5
and more exchange acidity than the sum of bases plus
KCl-extractable aluminum;
h. Have an ECEC (sum of bases KCl-extractable Al) of
more than 1.5 cmol(+) kg^{-1} clay in all subhorizons to a depth
of 150 cm below the soil surface; and
i. When neither irrigated nor fallowed to store moisture:
1. If the soil temperature regime is mesic or thermic, are
moist more than six-tenths of the time in half or more
years in some part of the moisture control section (not
necessarily the same part) when the soil temperature at a
depth of 50 cm exceeds 5°C; or
2. If the soil temperature regime is hyperthermic,
isomesic, or warmer, are moist in most years in some or
all parts of the moisture control section for 180 or more
days during a period when the soil temperature at a
depth of 50 cm exceeds 8°C; and
j. When neither irrigated nor fallowed to store moisture:
1. If the soil temperature regime is mesic or thermic, are
dry for more than 135 cumulative days in some part of
the moisture control section when the soil temperature at
a depth of 50 cm exceeds 5°C; or
2. If the soil temperature regime is hyperthermic,
isomesic, or warmer, the soils are dry in some or all parts
of the moisture control section for more than 90 days
during a period when the soil temperature at a depth of
50 cm exceeds 8°C;.
Acric Kanhaplustults are like Typic Kanhaplustults except
for *h*, with or without any of *c*, *d*, or *g*.
Andic Kanhaplustults are like Typic Kanhaplustults except
for *g*, with or without *c* or *d* or both.
Andic Udic Kanhaplustults are like Typic Kanhaplustults
except for *g* and *j*.
Aquic Kanhaplustults are like Typic Kanhaplustults except
for *b*.

Arenic Kanhaplustults are like Typic Kanhaplustults except for *e*.
Aridic Kanhaplustults are like Typic Kanhaplustults except for *i*.
Epiaquic Kanhaplustults are like Typic Kanhaplustults except for *c*.
Lithic Kanhaplustults are like Typic Kanhaplustults except for *a*.
Plinthic Kanhaplustults are like Typic Kanhaplustults except for *d*.
Rhodic Kanhaplustults are like Typic Kanhaplustults except for *f*.
Udic Kanhaplustults are like Typic Kanhaplustults except for *j*.

Paleustults

Paleustults are the Ustults that
1. Have a clay distribution such that the percentage of clay does not decrease from its maximum amount by >20 percent of that maximum within 150 cm of the soil surface, or the layer in which the clay percentage decreases by more than 20 percent has at least 5 percent of the volume consisting of skeletans on faces of peds and there is at least 3 percent (absolute) increase in clay below the layer.
2. Do not have plinthite that forms a continuous phase or constitutes more than half the matrix in any subhorizon within 150 cm of the soil surface; and
3. Do not have a fragipan or kandic horizon.

Plinthustults

Plinthustults are the Ustults that have plinthite that forms a continuous phase or constitutes more than half the matrix within some subhorizon in the upper 150 cm of the soil.

Rhodustults

Rhodustults are the Ustults that do not have a fragipan and also
1. Have an epipedon that has a color value, moist, less than 4 in all parts;
2. Have an argillic horizon that has a color value, dry, less than 5 in all subhorizons and not more than 1 unit higher than the value, moist;
3. Do not have plinthite that forms a continuous phase or constitutes more than half the matrix in any subhorizon in the upper 150 cm of the soil; and
4. Have a clay distribution with depth such that the percentage of clay decreases from its maximum amount by >20 percent of that maximum within 150 cm of the soil surface, and the layer in which the percentage of clay is less than the maximum has less than 5 percent of the volume consisting of skeletans on faces of peds or there is less than 3 percent (absolute) increase ion clay below the layer.
5. Do not have a fragipan or a kandic horizon.

Xerults

Key to great groups

FEA. Xerults that have an argillic horizon that has <10 percent weatherable minerals in the 20- to 200-micrometer fraction in its upper 50 cm and have a clay distribution such that the percentage of clay does not decrease from its maximum amount by >20 percent of that maximum within 1.5 m of the soil surface, or the layer in which the percentage of clay is less than the maximum has skeletans on ped faces or has 5 percent or more plinthite by volume.

Palexerults, p. 257

FEB. Other Xerults.

Haploxerults, p. 257

Haploxerults

Distinctions between Typic Haploxerults and other subgroups

Typic Haploxerults are the Haploxerults that
a. Do not have the following combination of characteristics in the upper 25 cm or more of the argillic horizon:
    (1) Mottles that have a color value, moist, or 4 or more and chroma, moist, of 2 or less and also mottles of higher chroma that are due to segregation of iron; and
    (2) Saturation with water in the mottled zone at some time of year when the soil temperature in that zone is 5°C or higher, or the soil is artificially drained;
b. Do not have a lithic contact within 50 cm of the mineral soil surface;
c. Have texture finer than loamy fine sand in some part of the argillic horizon and have an argillic horizon that does not have lamellae in at least its upper 25 cm;
d. Do not have a layer in the upper 75 cm that has texture finer than loamy fine sand, that is as much as 18 cm thick, that has bulk density (at 33 kPa water tension) of 0.95 g per cubic centimeter or less in the fine-earth fraction, and that has either of the following:
    (1) A ratio of measured clay to 1500 kPa water (percentages) of 1.25 or less; or
    (2) A ratio of CEC (at pH near 8) to 1500 kPa water of >1.5 and more exchange acidity than the sum of bases plus KCl-extractable aluminum;
e. Have a continuous argillic horizon throughout each pedon, not interrupted by ledges of bedrock; and
f. Do not have an epipedon as thick as 50 cm if its particle-size class is sandy throughout.
Ruptic-Lithic-Xerochreptic Haploxerults are like Typic Haploxerults except for *b* and *e*.

Palexerults

Palexerults are the Xerults that
1. Have both the following characteristics:
    a. An argillic horizon that in its upper 50 cm has <10 percent weatherable minerals in the 20- to 200-micrometer fraction; and

b. A clay distribution such that the percentage of clay does not decrease from its maximum amount by >20 percent of that maximum within 1.5 m of the soil surface, or the layer in which the percentage of clay is less than the maximum has skeletans on ped faces or has 5 percent or more plinthite by volume;

2. Have a color value, moist, of 4 or more in some part of the epipedon or have an argillic horizon that has a color value, dry, of 5 or more in some subhorizon or a color value, moist, of 4 or more;

3. Do not have plinthite that forms a continuous phase or constitutes more than half the matrix in any subhorizon within 1.25 m of the soil surface; and

4. Do not have a fragipan.

LITERATURE CITED

Sys, C. 1969. The soils of central Africa in the American classification - 7th approximation. African Soils. XIV:25-44.

Chapter 14
Vertisols

Key To Suborders

DA. Vertisols that have a thermic, mesic, or frigid soil temperature regime and, unless irrigated, have cracks that open and close once each year and remain open for 60 consecutive days or more in the 90 days following the summer solstice in more than 7 out of 10 years but that are closed for 60 consecutive days or more during the 90 days following the winter solstice.

Xererts, p. 261

DB. Other Vertisols that, unless irrigated, have in most years cracks that either remain open throughout the year or are closed for less than 60 consecutive days at a period when the soil temperature at a depth of 50 cm is continuously higher than 8°C.

Torrerts, p. 259

DC. Other Vertisols that have cracks that open and close one or more times during the year in most years but do not remain open for as many as 90 cumulative days in most years.

Uderts, p. 259

DD. Other Vertisols.

Usterts, p. 260

Torrerts

Distinctions between Typic Torrerts and other subgroups

Typic Torrerts are the Torrerts that
a. Have a color value, moist, of 4 or more in the surface horizon in more than half of each pedon, or the upper horizon that has a color value (moist) less than 4 is <30 cm thick; and
b. Do not have prismatic or blocky structure accompanied by clay skins on ped faces that have a color value lower than that in the matrix within 1 m of the soil surface.
Mollic Torrerts are like Typic Torrerts except for *a.*
Paleustollic Torrerts are like Typic Torrerts except for *a* and *b.*

Uderts

Key to great groups

DCA. Uderts that have a chroma, moist, of 1.5 or more dominant in the matrix of some subhorizon in the upper 30 cm in more than half of each pedon.

Chromuderts, p. 260

DCB. Other Uderts.

Pelluderts, p. 260

Chromuderts

Distinctions between Typic Chromuderts and other subgroups

Typic Chromuderts are the Chromuderts that
a. Do not have distinct or prominent mottles within 50 cm
of the soil surface in more than half of each pedon (the
terms refer to contrast, not to size of the mottles); and
b. Have a color value, moist, less than 3.5 and a value, dry,
less than 5.5 throughout the upper 30 cm in more than half
of each pedon.
Aquentic Chromuderts are like Typic Chromuderts except for
a and *b.*
Aquic Chromuderts are like Typic Chromuderts except for *a.*
Entic Chromuderts are like Typic Chromuderts except for *b.*

Pelluderts

Typic Pelluderts are the Pelluderts that have a color value,
moist, less than 3.5 and a value, dry, less than 5.5 throughout the upper 30 cm in more than half of each pedon.

Description of subgroups

Typic Pelluderts.--The central concept or typic subgroup of
Pelluderts is fixed on soils that have a thick, dark epipedon.
The color of the surface layer and the content of organic
matter are about the only visible properties that vary in the
great group. Base saturation also varies. These soils are
extensive locally on the Gulf Coast in Texas. They were
formerly cultivated, but most of them are now used for
pasture.

Entic Pelluderts.--These soils are like Typic Pelluderts except for their color value. This is the only subgroup besides
Typic Pelluderts that has been recognized in the United
States. A surface horizon that has a color value, moist, less
than 3.5 and a color value, dry, less than 5.5 either is absent
or is <30 cm thick in more than half of each pedon. These
soils are not extensive in the United States. They are used
mostly for pasture.

Usterts

Key to great groups

DDA. Usterts that have a chroma, moist, of 1.5 or more in
some part of the matrix of the upper 30 cm in more than
half of each pedon.

Chromusterts, p. 261

DDB. Other Usterts.

Pellusterts, p. 261

Chromusterts

Distinctions between Typic Chromusterts and other subgroups

Typic Chromusterts are the Chromusterts that
a. Have a color value, moist, less than 3.5 and a value, dry, less than 5.5 throughout the upper 30 cm or more in more than half of each pedon;
b. Do not have, within 1 m of the soil surface, prismatic or blocky structure accompanied by clay skins on ped faces that have a color value lower than that in the matrix; and
c. Have cracks that remain open more than 150 cumulative days in most years and have a mean annual soil temperature that is 15°C or higher.
Entic Chromusterts are like Typic Chromusterts except for *a*.
Paleustollic Chromusterts are like Typic Chromusterts except for *b* or for *a* and *b*.
Udic Chromusterts are like Typic Chromusterts except for *c* and have cracks that remain open from 90 to 150 cumulative days in most years.
Udorthentic Chromusterts are like Typic Chromusterts except for *a* and *c* and have cracks that remain open from 90 to 150 cumulative days in most years.

Pellusterts

Distinctions between Typic Pellusterts and other subgroups

Typic Pellusterts are the Pellusterts that
a. Have a color value, moist, less than 3.5 and a value, dry, less than 5.5 throughout the upper 30 cm in more than half of each pedon;
b. Have cracks that remain open for more than 150 cumulative days during each year and have a mean annual soil temperature that is 15°C or higher; and
c. Do not have within 1 m of the soil surface prismatic or blocky structure accompanied by clay skins on ped faces that have a color value lower than that in the matrix.
Entic Pellusterts are like Typic Pellusterts except for *a*.
Udic Pellusterts are like Typic Pellusterts except for *b*.
Udorthentic Pellusterts are like Typic Pellusterts except for *a* and *b*.

Xererts

Key to great groups

DAA. Xererts that have a dominant chroma, moist, of 1.5 or more in the matrix of some subhorizon in the upper 30 cm in more than half of each pedon.

Chromoxererts, p. 262

DAB. Other Xererts.

Pelloxererts, p. 262

Chromoxererts

Distinctions between Typic Chromoxererts and other subgroups

Typic Chromoxererts are the Chromoxererts that
a. Do not have distinct or prominent mottles (these terms
refer to contrast, not size) within 50 cm of the soil surface
in more than half of each pedon;
b. Have a color value, moist, less than 3.5 and a value, dry,
less than 5.5 throughout the upper soil to a depth of 30 cm
in more than half of each pedon; and
c. Do not have, within 1 m of the soil surface, prismatic or
blocky structure accompanied by clay skins on ped faces
that have a color value lower than that in the matrix.
Aquic Chromoxererts are like Typic Chromoxererts except
for *a.*
Entic Chromoxererts are like Typic Chromoxererts except
for *b.*
Palexerollic Chromoxererts are like Typic Chromoxererts
except for *c.*

Pelloxererts

Distinctions between Typic Pelloxererts and other subgroups

Typic Pelloxererts are the Pelloxererts that
a. Have in all subhorizons to a depth of 1 m a chroma, both
dry and moist, less than 1.5 or, if the chroma is 1.5 or
higher, there are in some subhorizon between 30 cm and 1
m distinct or prominent mottles, or concretions that are due
to segregated iron or manganese; and
b. Have a color value, moist, less than 3.5 and a value, dry,
less than 5.5 throughout the upper 30 cm in more than half
of each pedon.
Chromic Pelloxererts are like Typic Pelloxererts except for
a.
Entic Pelloxererts are like Typic Pelloxererts except for *b.*

Index

SI Units Conversion Table

The following conversions have been made in the text of the *Keys*.

CEC:

1 meq/100 g soil = 1 cmol(+) kg^{-1} soil

Conductivity:

1 mmho/cm = 1 dS m^{-1}

Pressure:

15-bar water = 1500 kPa water

1/3-bar water = 33 kPa water

Concentration:

1N = 1*M*

Length:

1 micron = 1 micrometer

DESIGNATIONS FOR HORIZONS AND LAYERS

Genetic horizons are not the equivalent of the diagnostic horizons of *Soil Taxonomy*. Designations of genetic horizons express a qualitative judgment about the kind of changes that are believed to have taken place. Diagnostic horizons are quantitatively defined features used to differentiate between taxa. The diagnostic horizons may encompass several genetic horizons, and changes implied by genetic horizon designations may not be large enough to justify recognizing different diagnostic criteria.

MASTER HORIZONS AND LAYERS

The capital letters **O, A, E, B, C,** and **R** represent the master horizons and layers of soils. The capital letters are the base symbols to which other characters are added to complete the designations. Most horizons and layers are given a single capital letter symbol; some require two.

O horizons or layers: *Layers dominated by organic material, except limnic layers[1] that are organic. Some are saturated with water for long periods or were once saturated but are now artificially drained; others have never been saturated.*

Some O layers consist of undecomposed or partially decomposed litter, such as leaves, needles, twigs, moss, and lichens, that has been deposited on the surface; they may be on top of either mineral or organic soils. Other O layers, called peat, muck, or mucky peat, are organic material that was deposited underwater and that has decomposed to varying stages. The mineral fraction of such material is only a small percentage of the volume of the material and generally is much less than half of the weight. Some soils consist entirely of material designated as O horizons or layers.

An O layer may be on the surface of a mineral soil or at any depth beneath the surface if it is buried. A horizon formed by illuviation of organic material into a mineral subsoil is not an O horizon, though some horizons formed in this manner contain much organic matter.

A limnic layer that is organic is designated a C layer.

A horizons: *Mineral horizons that formed at the surface or below an O horizon and*
> *(1) are characterized by an accumulation of humified organic matter intimately mixed with the mineral fraction and not dominated by properties characteristic of E or B horizons (defined below) or*
> *(2) have properties resulting from cultivation, pasturing, or similar kinds of disturbance.*

If a surface horizon has properties of both A and E horizons but the feature emphasized is an accumulation of humified organic matter, it is designated an A

---

[1] Coprogenous earth, diatomaceous earth, marl.

horizon. In some places, as in warm arid climates, the undisturbed surface horizon is less dark than the adjacent underlying horizon and contains only small amounts of organic matter. It has a morphology distinct from the C layer, though the mineral fraction is unaltered or only slightly altered by weathering. Such a horizon is designated A because it is at the surface. However, recent alluvial or eolian deposits that retain fine stratification are not considered to be an A horizon unless cultivated.

E horizons: *Mineral horizons in which the main feature is loss of silicate clay, iron, aluminum, or some combination of these, leaving a concentration of sand and silt particles of quartz or other resistant minerals.*

An E horizon is usually, but not necessarily, lighter in color than an underlying B horizon. In some soils the color is that of the sand and silt particles, but in many soils coats of iron or other compounds mask the color of the primary particles. An E horizon is most commonly differentiated from an overlying A horizon by lighter color and generally has measurably less organic matter than the A horizon. An E horizon is most commonly differentiated from an underlying B horizon in the same sequum by color of higher value or lower chroma, by coarser texture, or by a combination of these properties. An E horizon is commonly near the surface below an O or A horizon and above a B horizon, but the symbol E may be used without regard to position in the profile for any horizon that meets the requirements and that has resulted from soil genesis.

B horizons: *Horizons that formed below an A, E, or O horizon and are dominated by obliteration of all or much of the original rock structure[2] and by*
(1) illuvial concentration of silicate clay,
iron, aluminum, humus, carbonates, gypsum, or silica,
alone or in combination;
(2) evidence of removal of carbonates;
(3) residual concentration of sesquioxides;
(4) coatings of sesquioxides that make the horizon
conspicuously lower in value, higher in chroma, or redder in hue than overlying and underlying horizons
without apparent illuviation of iron;
(5) alteration that forms silicate clay or liberates oxides or both and that forms granular, blocky, or prismatic structure if volume changes accompany changes in moisture content; or
(6) any combination of these.

Obviously there are several kinds of B horizon. No common location within the soil characterizes them, but all are subsurface horizons or were originally. Included as B horizons where contiguous to another

---

[2] Rock structure includes fine stratification in unconsolidated or weakly consolidated sediment or pseudomorphs of weathered minerals retaining their positions relative to each other and to unweathered minerals in saprolite from consolidated rocks.

genetic horizon are layers of illuvial concentration of carbonates, gypsum, or silica that are the result of pedogenic processes (these layers may or may not be cemented) and brittle layers that have other evidence of alteration, such as prismatic structure or illuvial accumulation of clay.

Examples of layers that are not B horizons are layers in which clay films coat rock fragments or are on finely stratified unconsolidated sediments, whether the films were formed in place or by illuviation, and layers into which carbonates have been illuviated unless contiguous to an overlying genetic horizon.

C horizons or layers: *Horizons or layers, excluding hard bedrock, that are little affected by pedogenic processes and lack properties of O, A, E, or B horizons. Most are mineral layers, but limnic layers[3], whether organic or inorganic, are included. The material of C layers may be either like or unlike that from which the solum presumably formed. A C horizon may have been modified even if there is no evidence of pedogenesis.*

Included as C layers are sediments, saprolite, and consolidated bedrock that when moist can be dug with a spade. Some soils form in material that is already highly weathered, and such material that does not meet the requirements of A, E, or B horizons is designated C. Changes not considered pedogenic are those not related to overlying horizons. Layers having accumulations of silica, carbonates, or gypsum or more soluble salts are included in C horizons, even if indurated, unless these layers are obviously affected by pedogenic processes; then they are a B horizon.

R Layers: Hard Bedrock.

Granite, basalt, quartzite, and indurated limestone or sandstone are examples of bedrock that are designated R. The bedrock of an R layer is sufficiently coherent when moist to make hand digging with a spade impractical, although it may be chipped or scraped with a spade. Some R layers can be ripped with heavy power equipment. The bedrock may contain cracks, but these are few enough and small enough that few roots can penetrate. The cracks may be coated or filled with clay or other material.

Transitional horizons

There are two kinds of transitional horizon. In one, the properties of an underlying or overlying horizon are superimposed on properties of the other horizon throughout the transition zone. In the other, parts that are characteristic of an overlying or underlying horizon are enclosed by parts that are characteristic of the other horizon. Special conventions are used to designate these kinds of horizons.

---

[3] Coprogenous earth, diatomaceous earth, marl.

*Horizons dominated by properties of one master horizon
but having subordinate properties of another.* Two capital
letter symbols are used, as AB, EB, BE, BC. The master
horizon symbol that is given first designates the kind of
horizon whose properties dominate the transitional horizon.
An AB horizon, for example, has characteristics of both an
overlying A horizon and an underlying B horizon, but is
more like the A than like the B.

In some cases, a horizon can be designated as transitional
even if one of the master horizons to which it is appar-
ently transitional is not present. A BE horizon may be recognized in a truncated soil if its properties are similar to
those of a BE horizon in a soil in which the overlying E
horizon has not been removed by erosion. An AB or a BA
horizon may be recognized where bedrock underlies the
transitional horizon. A BC horizon may be recognized even
if no underlying C horizon is present; it is transitional to
assumed parent material.

*Horizons in which distinct parts have recognizable proper-
ties of the two kinds of master horizons indicated by the
capital letters.* The two capital letters are separated by a
virgule (/), as E/B, B/E, B/C. Most of the individual parts
of at least one of the components are surrounded by the
other.

The designation may be used even though horizons similar
to one or both of the components are not present, if the
separate components can be recognized in the transitional
horizon. The first symbol is that of the horizon that makes
up the greater volume.

SUBORDINATE DISTINCTIONS WITHIN MASTER
HORIZONS AND LAYERS

Lower case letters are used as suffixes to designate specific kinds of master horizons and layers. The symbols and
their meanings are as follows:

The word "accumulation" is used in many of the definitions.
As used here accumulation means that the horizon must have
more of the material in question than the parent material is
presumed to have had.

    a   *Highly decomposed organic material*

        This symbol is used with "O" to indicate the most
        highly decomposed of the organic materials.
        Rubbed fiber content averages less than about 1/6
        of the volume.

    b   *Buried genetic horizon*

        This symbol is used in mineral soils to indicate
        identifiable buried genetic horizons if the major
        features of the buried horizon had been established
        before it was buried. It is not used in organic soils
        or to separate an organic layer from a mineral
        layer. Genetic horizons may or may not have
        formed in the overlying material, which may be

either like or unlike the assumed parent material of
the buried soil.

c   *Concretions or hard nonconcretionary nodules*

This symbol is used to indicate a significant accumulation of concretions or nonconcretionary nodules cemented by material other than silica. This
symbol is not used if concretions or nodules are
dolomite or calcite or more soluble salts, but it
is used if the nodules or concretions are iron, aluminum, manganese, or titanium. Their consistence
is specified in the horizon description.

d   *Dense unconsolidated sediments or materials*

This symbol is used to indicate naturally occurring
or man-made, unconsolidated sediments or materials with high bulk density such as dense basal till,
plow pans and other mechanically compacted zones.
The layer is root restrictive and roots do not enter
except along fracture planes.

e   *Organic material of intermediate decomposition*

This symbol is used with "O" to indicate organic
materials of intermediate decomposition. Rubbed
fiber content is 1/6 to 2/5 of the volume.

f   *Frozen soil*

This symbol is used to indicate that the horizon or
layer contains permanent ice. Symbol is not used
for seasonally frozen layers or for "dry permafrost"
(material that is colder than 0 °C but does not contain ice).

g   *Strong gleying*

This symbol is used to indicate either that iron has
been reduced and removed during soil formation or
that saturation with stagnant water has preserved a
reduced state. Most of the affected layers have low
chroma and many are mottled. The low chroma
can be the color of reduced iron or the color of
uncoated sand and silt particles from which iron
has been removed. Symbol "g" is not used for soil
materials of low chroma, such as some shales or E
horizons, unless they have a history of wetness.
If "g" is used with "B", pedogenic change in addition to gleying is implied. If no other change has
taken place, the horizon is designated Cg.

h   *Illuvial accumulation of organic matter*

This symbol is used with "B" to indicate the accumulation of illuvial, amorphous, dispersible organic
matter-sesquioxide complexes if the sesquioxide
component is dominated by aluminum but is present only in very small quantities.
The organosesquioxide material coats sand and silt

particles or may occur as discrete pellets. In some horizons, coatings have coalesced, filled pores, and cemented the horizon. The symbol "h" is also used in combination with "s" as "Bhs" if the amount of sesquioxide component is significant but value and chroma of the horizon are 3 or less.

i   *Slightly decomposed organic material*

This symbol is used with "O" to indicate the least decomposed of the organic materials. Rubbed fiber content is more than about 2/5 of the volume.

k   *Accumulation of carbonates*

This symbol is used to indicate accumulation of alkaline earth carbonates, commonly calcium carbonate.

m   *Cementation or induration*

This symbol is used to indicate continuous or nearly continuous cementation. Symbol is used only for horizons that are more than 90 percent cemented, though they may be fractured. Roots penetrate "m" horizons only through cracks. The cementing material is also symbolized. If 90 percent or more of the horizon is cemented by carbonates, "km" is used; by silica, "qm"; by iron, "sm"; by gypsum, "ym"; by both lime and silica, "kqm"; by salts more soluble than gypsum, "zm".

n   *Accumulation of sodium*

This symbol is used to indicate accumulation of exchangeable sodium.

o   *Residual accumulation of sesquioxides*

This symbol is used to indicate residual accumulation of sesquioxides.

p   *Plowing or other disturbance*

This symbol is used to indicate disturbance of the surface layer by cultivation, pasturing, or similar uses. A disturbed organic horizon is designated Op. A disturbed mineral horizon, even though clearly once a E, B, or C horizon, is designated Ap.

q   *Accumulation of silica*

This symbol is used to indicate accumulation of secondary silica. If silica cements the layer and cementation is continuous or nearly continuous, "qm" is used.

r   *Weathered or soft bedrock*

This symbol is used with "C" to indicate layers of soft bedrock or saprolite, such as weathered igneous rock; partly consolidated soft sandstone, siltstone, or shale; roots cannot enter except along fracture planes. The material can be dug with a spade.

s   *Illuvial accumulation of sesquioxides and organic matter*

This symbol is used with "B" to indicate the accumulation of illuvial, amorphous, dispersable organic matter-sesquioxide complexes if both the organic matter and sesquioxide components are significant and the value and chroma of the horizon is more than 3. The symbol is also used in combination with "h" as "Bhs" if both the organic matter and sesquioxide components are significant and the value and chroma are 3 or less.

t   *Accumulation of silicate clay*

This symbol is used to indicate an accumulation of silicate clay that either has formed in the horizon or has been moved into it by illuviation. The clay can be in the form of coatings on ped surfaces or in pores, lamellae, or bridges between mineral grains.

v   *Plinthite*

This symbol is used to indicate the presence of iron-rich, humus-poor, reddish material that is firm or very firm when moist and that hardens irreversibly when exposed to the atmosphere and to repeated wetting and drying. These properties are characteristic of plinthite.

w   *Development of color or structure*

This symbol is used with "B" to indicate development of color or structure, or both, with little or no apparent illuvial accumulation of material. It should not be used as a substitute for a transitional horizon.

x   *Fragipan character*

This symbol is used to indicate genetically developed firmness, brittleness, or high bulk density. These features are characteristic of fragipans, but some horizons designated "x" do not have all properties of a fragipan.

y   *Accumulation of gypsum*

This symbol is used to indicate accumulation of gypsum.

z   *Accumulation of salts more soluble than gypsum*

This symbol is used to indicate accumulation of salts more soluble than gypsum.

Conventions for using letter suffixes.--Many master horizons and layers that are symbolized by a single capital letter will have one or more lower case letter suffixes. Common exceptions are an undisturbed A horizon and many, if not most, E and C horizons and layers. Seldom are more than three suffixes needed.

When letter suffixes are used, they immediately follow the capital letter. If a surface horizon is disturbed, only "p" is used except where there are surface accumulations of $CaCO_3$, $CaSO_4$ or more soluble salts.

When more than one suffix is needed, the following letters, if used, are written first: a, e, i, h, r, s, t, and w. Except for Bhs or Crt[4] horizons, none of these letters are used in combination in a single horizon.

A horizon is never designated Bth, Bts, or Btw, though a Bw, Bs, or Bh horizon may be above or below a Bt horizon. A B horizon that has a significant accumulation of clay and also shows evidence of development of color or structure, or both, is designated Bt, ("t" has precedence over "w", "s", and "h"). A B horizon that is gleyed or that has accumulations of carbonates, silica, gypsum, salts more soluble than gypsum, or residual accumulation of sesquioxides carries the appropriate symbol--g, k, q, y, z, or o. If illuvial clay is also present, "t" precedes the other symbol: Bto. Suffixes "h", "s", and "w" are not used with g, k, q, y, z, or o.

If a horizon is buried, the suffix "b" is written last. Suffix "b" is used only for buried mineral soils. If more than one suffix is needed and the horizon is not buried, these symbols, if used, are written last: c, f, g, m, and x. Some examples: Btc, Ckm, and Bsv.

Lower case letter suffixes are not used with transitional horizons unless needed for explanatory purposes; for example, use of "k" is appropriate in the sequence A-ACk1-ACk2-AC-C to indicate an accumulation of carbonates in the upper parts of the AC horizon.

Vertical subdivision.--Commonly a horizon or layer designated by a single combination of letters needs to be subdivided. The Arabic numerals used for this purpose always follow all letters. Within a C, for example, successive layers could be C1, C2, C3, etc.; or if the lower part is gleyed and the upper part is not, the designations could be C1-C2-Cg1-Cg2 or C-Cg1-Cg2-R.

These conventions apply whatever the purpose of subdivision. In many soils, horizons that would be identified by one unique set of letters are subdivided on the basis of evident morphological features, such as structure, color, or texture. These divisions are numbered consecutively. The numbering starts with 1 at whatever level in the profile any element of the letter symbol changes. Thus Bt1-Bt2-Btk1-Btk2 is used, not Bt1-Bt2-Btk3-Btk4. The numbering of vertical subdivisions within a horizon is not interrupted at a discontinuity (indicated by a numerical prefix) if the same

---

[4] Indicating weathered bedrock or saprolite in which clay skins are present.

letter combination is used in both materials: Bs1-Bs2-2Bs3-2Bs4 is used, not Bs1-Bs2-2Bs1-2Bs2.

Sometimes, thick layers are subdivided during sampling for laboratory analyses even though differences in morphology are not evident in the field. These layers need to be identified, and this is done simply by numbering each subdivision consecutively within a layer having a unique symbol, starting at the top. For example, four layers of a Bt horizon sampled by 10-cm increments would be designated Bt1, Bt2, Bt3, Bt4.

Discontinuities.--In mineral soils Arabic numerals are used as prefixes to indicate discontinuities. Wherever needed, they are used preceding A, E, B, C, and R. These prefixes are distinct from Arabic numerals used as suffixes to denote vertical subdivisions.

A discontinuity is a significant change in particle-size distribution or mineralogy that indicates a difference in the material from which the horizons formed or, except for some buried soils, a significant difference in age. Symbols to identify discontinuities are used only when they will contribute substantially to the reader's understanding of relationships among horizons. The significance of a given kind of discontinuity may be large in one soil and small in another, or even large in one horizon and small in another of the same profile. Stratification common to soils formed in alluvium is not designated as discontinuities even if particle size distribution differs markedly from layer to layer unless genetic horizons have formed in the contrasting layers.

Where a soil has formed entirely in one kind of material, a prefix is omitted from the symbol; the whole profile is material 1. Similarly, the uppermost material in a profile having two or more contrasting materials is understood to be material 1, but the number is omitted. Numbering starts with the second layer of contrasting material, which is designated "2". Underlying contrasting layers are numbered consecutively. Even though a layer below material 2 is similar to material 1, it is designated "3" in the sequence. The numbers indicate a change in the material, not the type of material. Where two or more consecutive horizons formed in one kind of material, the same prefix number is applied to all of the horizon designations in that material: Ap-E-Bt1-2Bt2-2Bt3-2BC. The number suffixes designating subdivisions of the Bt horizon continue in consecutive order across the discontinuity.

If an R layer is below a soil that formed in residuum and the material of the R layer is presumed to be like that from which the material of the soil weathered, the Arabic number prefix is not used. If the R layer would not produce material like that in the solum, the number prefix is used, as in A-Bt-C-2R or A-Bt-2R. If part of the solum formed in residuum, "R" is given the appropriate prefix: Ap-Bt1-2Bt2-2Bt3-2C1-2C2-2R.

Buried horizons (designated "b") are special problems. A buried horizon is obviously not in the same deposit as hori-

zons in the overlying deposit. Some buried horizons, however, formed in material lithologically like that of the overlying deposit. A prefix is not used to distinguish material of
such buried horizons. If the material in which a horizon of
a buried soil formed is lithologically unlike that of the
overlying material, the discontinuity is designated by number prefixes and the symbol for a buried horizon is used as
well: Ap-Bt1-Bt2-BC-C- 2Ab-2Btb1-2Btb2-2C.

In organic soils, discontinuities between different kinds
of layers are not identified. In most cases the differences
are shown by the letter suffix designations, if the different
layers are organic, or by the master symbol if the different
layers are mineral.

Use of the prime.--Identical designations may be appropriate for two or more horizons or layers separated by at least
one horizon or layer of a different kind in the same pedon.
The sequence A-E-Bt-E-Btx-C is an example: the soil has
two E horizons. To make communication easier, the prime
is used with the master horizon symbol of the lower of two
horizons having identical letter designations: A-E-Bt-E'-
Btx-C. The prime is applied to the capital letter designation, and any lower case symbols follow it: B't. The prime
is not used unless all letters of the designations of two different layers are identical. Rarely, three layers have identical letter symbols; a double prime can be used: E".

The same principle applies in designating layers of organic soils. The prime is used only to distinguish two or
more horizons that have identical symbols: Oi-C-O'i-C' or
Oi-C-Oe-C'. The prime is added to the lower C layer to
differentiate it from the upper.